WITHDRAWN

D0859941

Current Topics in Microbiology 208 and Immunology

Editors

A. Capron, Lille · R.W. Compans, Atlanta/Georgia
M. Cooper, Birmingham/Alabama · H. Koprowski,
Philadelphia/Pennsylvania · I. McConnell, Edinburgh
F. Melchers, Basel · M. Oldstone, La Jolla/California
S. Olsnes, Oslo · M. Potter, Bethesda/Maryland
H. Saedler, Cologne · P.K. Vogt, La Jolla/California
H. Wagner, Munich · I. Wilson, La Jolla/California

Springer
Berlin
Heidelberg
New York
Barcelona
Budapest
Hong Kong
London
Milan
Paris
Santa Clara
Singapore
Tokyo

Transcriptional Control of Cell Growth:

The E2F Gene Family

Edited by P.J. Farnham

With 17 Figures

Springer

Dr. PEGGY J. FARNHAM
Associate Professor of Oncology

McArdle Laboratory for Cancer Research
University of Wisconsin Medical School
1400 University Avenue
Madison, WI 53706
USA

Cover illustration: Courtesy of Erika Meyer, Department of Oncology, and Christopher Lukas, Department of Computer Science, University of Wisconsin at Madison, USA. Shown is a schematic representation of the four stages (G1, S, G2, and M) of the proliferative cell cycle. Within the circle is an encoded stereogram representing heterodimeric E2F and DP proteins binding to DNA (a black and white version of the decoded stereogram is also provided).

Cover design: Künkel+Lopka, Ilvesheim

ISSN 0070-217X
ISBN 3-540-60113-9 Springer-Verlag Berlin Heidelberg New York

This work is subject to copyright. All rights are reserved, whether the whole or part of the material is concerned, specifically the rights of translation, reprinting, reuse of illustrations, recitation, broadcasting, reproduction on microfilm or in any other way, and storage in data banks. Duplication of this publication or parts thereof is permitted only under the provisions of the German Copyright Law of September 9, 1965, in its current version, and permission for use must always be obtained from Springer-Verlag. Violations are liable for prosecution under the German Copyright Law.

© Springer-Verlag Berlin Heidelberg 1996
Library of Congress Catalog Card Number 15-12910
Printed in Germany

The use of general descriptive names, registered names, trademarks, etc. in this publication does not imply, even in the absence of a specific statement, that such names are exempt from the relevant protective laws and regulations and therefore free for general use.

Product liability: The publishers cannot guarantee the accuracy of any information about dosage and application contained in this book. In every individual case the user must check such information by consulting other relevant literature.

Typesetting: Thomson Press (India) Ltd, Madras
SPIN: 10495401   27/3020/SPS – 5 4 3 2 1 0 – Printed on acid-free paper

Preface

It is of critical importance to maintain an appropriate balance between proliferation and quiescence or differentiation throughout the lifespan of all animals. An important control point in this balance occurs in the G_1 phase of the cell cycle. On the basis of environmental cues a cell in G_1 must decide whether to continue through the proliferative cycle and enter S phase (where DNA replication occurs) or to exit from the proliferative cycle into a nonreplicating state. Alterations in the mechanisms that normally control this decision can lead to cancer, cell death, or loss of differentiated cellular phenotypes. The identification of the E2F gene family of transcription factors has allowed a more complete understanding of how the cell maintains an appropriate proliferative state. This volume provides an up-to-date account of present reports concerning E2F as well as a framework for future investigations.

E2F activity requires heterodimerization of two partners. Either partner can be one of several different transcription factors; E2F1, E2F2, E2F3, E2F4, or E2F5 can heterodimerize with either DP1 or DP2. Cellular promoters whose E2F sites mediate a link between transcription and proliferation drive genes whose products are required for DNA synthesis and genes that encode regulators of cell growth. A detailed analysis of the role that E2F family members play in transcription from these promoters is presented in the chapter by J.E. SLANSKY and P.J. FARNHAM.

Investigations of E2F protein complexes have aided greatly in the refinement of models of both negative and positive cell growth control. For example, although it was known that loss of the retinoblastoma (Rb) protein is associated with human neoplasias, it was not clear how this tumor suppressor protein functions. It has since been discovered that Rb can bind to and inactivate E2F. Although cyclins were known to be the growth-regulated component of kinase complexes, it was not clear what proteins are targets of these kinases. We now know that several G_1 cyclins can phosphorylate Rb and release it from interaction with E2F. These interactions of E2F with cyclins and Rb are described in the chapter by D. COBRINIK.

Several different viral oncoproteins function by releasing E2F from inhibitory complexes to produce an environment favourable for viral DNA replication; the involvement of E2F in viral oncogenesis is described in the chapter by W.D. Cress and J.R. Nevins. Increased levels of E2F can also circumvent a cell's normal response to environment cues and have severe consequences such as neoplastic transformation and cell death. The effects of overexpression of various E2F family members is described in the chapter by P. ADAMS and W.G. KAELIN.

Although much has been learned about the E2F gene family in the last several years, we still lack a complete understanding of the functions of individual family members. The chapter by L. Breeden summarizes the mechanisms by which yeast control the G1 to S phase transition, in the hope that parallels can be drawn that will provide insight into mammalian cell growth control.

Madison   P.J. FARNHAM

List of Contents

List of Contributors

(Their addresses can be found at the beginning of their respective chapters.)

Introduction to the E2F Family:
Protein Structure and Gene Regulation

J.E. Slansky and P.J. Farnham

1 Identification and Cloning of E2F Family Members

The E2F gene family was identified through several different types of investigations, including studies of viral oncogenes, positive and negative growth regulatory proteins, and cellular promoters. As described below and by Cobrinik, Cress and Nevins, and Adams and Kaelin (this volume), the E2F family (which includes at least seven members) has not only been implicated in controlling viral and cellular gene expression, but also is thought to play an important role in growth regulatory processes such as the G_1 to S phase transition and programmed cell

McArdle Laboratory for Cancer Research, University of Wisconsin Medical School, 1400 University Avenue, Madison, WI 53706, USA death. To understand how this gene family can mediate these important events it is first necessary to examine the structure and regulation of the individual family members.

1.1 E2F Genes

In 1986 investigators identified a cellular-protein activity required for transactivation of the adenovirus *E2* promoter by the E1A oncoprotein; this cellular DNA-binding activity was termed E2F (for E2 factor; KOVESDI et al. 1986). Further investigation indicated that transcriptional activation by E1A results from release of E2F from complexes with other cellular proteins (see CRESS and NEVINS this volume). Free E2F then interacts with the viral E4 protein which mediates cooperative protein binding at two E2F sites in the *E2* promoter (BAGCHI et al. 1990). Several viruses have evolved strategies to break up E2F-protein complexes (CHELLAPPAN et al. 1992). The domains of the viral oncoproteins that are required to disrupt E2F complexes are also required for viral transformation; thus increasing the amount of free E2F is thought to be essential for neoplastic transformation by viruses such as adenovirus, papilloma virus, and simian virus 40 (SV40).

Further understanding of how disruption of E2F complexes leads to tumorigenicity was provided by investigations with the retinoblastoma (Rb) tumor suppressor protein. As a tumor suppressor protein, Rb negatively regulates cell growth and, as a result, is a target of some viral oncogenes. Rb was discovered to be one of several proteins that is bound by the adenoviral E1A protein (WHYTE et al. 1988). Since it was known that E1A releases E2F from protein complexes, it was hypothesized that E2F-mediated transcription is activated through sequestration of Rb by E1A. Evidence to support this model came in 1991 when investigators showed that Rb, when bound to a column, can associate with cellular proteins that interact with an E2F-binding site (CHITTENDEN et al. 1991b). The association of Rb with E2F was confirmed by gel mobility shift assays (BANDARA et al. 1991; CHELLAPPAN et al. 1991). Rb negatively regulates progression through G_1 into S phase (GOODRICH et al. 1991) and mutations in Rb that eliminate its ability to impede cell growth also eliminate its ability to bind to E2F (QIN et al. 1992). Additional experiments have demonstrated that Rb can directly repress E2F-mediated transcription (DYNLACHT et al. 1994b). Thus, it is believed that when viral oncoproteins disrupt E2F/Rb interactions, E2F is relesed to transactivate genes that are critical for progression through G_1 into S phase. Details concerning these target genes can be found in Sect. 4.

Based on the hypothesis that E2F and Rb interact directly, Rb protein was used to probe human cDNA expression libraries (HELIN et al. 1992; KAELIN et al. 1992; SHAN et al. 1992). Although three groups of investigators independently isolated multiple clones, sequence analysis indicated that each group had isolated a common cDNA. The identity of this gene (originally called *RBP3*, *RBAP-1*, or *Ap12*) as an E2F family member was confirmed by demonstrating that the encoded protein can bind to and transactivate a promoter containing E2F sites. Because several lines of investigations suggested the existence of more than one family member, the first clone was named *E2F1*. Two additional E2F family members were isolated by screening human cDNA libraries with an *E2F1* probe. These efforts identified human *E2F2* and *E2F3* (IVEY-HOYLE et al. 1993; LEES et al. 1993); an exhaustive search using this method did not detect any other E2Fs. A 3T3 cDNA library was then screened with the corresponding human clones to obtain mouse *E2F1* and *E2F3* (Fry et al., unpublished data; LI et al. 1994).

The inability of E2F1-, E2F2-, or E2F3-specific antibodies to react with an abundant E2F activity in gel mobility shift assays suggested that yet an additional E2F protein existed (CHITTENDEN et al. 1993; KAELIN et al. 1992; WU et al. 1995). None of the other clones isolated using Rb as a probe encode proteins that can bind to E2F sites. However, investigations of the different cellular complexes that bind to E2F sites indicated that proteins similar to, but distinct from, Rb can interact with E2F (see COBRINIK this volume). For example, p107, a cellular protein that has homology to Rb in the domain that binds to E2F (EWEN et al. 1992), was in E2F-specific gel shift complexes. Also, an E2F protein distinct from E2F1 was immunoprecipitated using p107-specific antibodies (DYSON et al. 1993). Therefore, others screened a cDNA expression library using p107, instead of Rb, as a probe and isolated a different family member, *E2F4* (BEIJERSBERGEN et al. 1994). *E2F4* was also cloned using polymerase chain reaction techniques with degenerate primers derived from E2F1 sequences (GINSBERG et al. 1994). E2F5, which binds specifically to another Rb-related protein, p130, has recently been cloned (R. Bernards, personal communication).

Genomic DNA has been cloned that corresponds to mouse *E2F1* and *E2F2* (M. Greenberg, personal communication; HSIAO et al. 1994) and human *E2F1*, *E2F2* and *E2F3* (JOHNSON et al. 1994b; LEES et al. 1993; NEUMAN et al. 1994). Mouse *E2F1* was mapped to chromosome 2 using multipoint linkage analysis which took advantage of a GT repeat in the 3' untranslated region of the *E2F1* mRNA (LI et al. 1994). Fluorescence in situ hybridization was used to localize human *E2F1* to chromosome 20q11, telomeric to the p107 locus (SAITO et al. 1995). The *E2F1* chromosomal assignment was confirmed by hybridizing a cDNA probe to Southern blots containing DNA extracted from a panel of rodent plus human hybrid cell lines containing different combinations of human chromosomes. Human *E2F2* and *E2F3* were mapped by in situ hybridization to metaphase chromosomes; *E2F2* is found at 1p36 and *E2F3* at 6q22 (LEES et al. 1993). The location of *E2F2* and *E2F3* on chromosomes 1 and 6, respectively, was also confirmed using a panel of rodent plus human hybrid cell lines. *E2F3* pseudogenes were also identified at 17q11–12 and 2q33–35 (LEES et al. 1993). Human *E2F4* is located at 16q22 as shown by in situ hybridization to metaphase chromosomes (GINSBERG et al. 1994). The alleles for the *E2F4* gene are polymorphic in humans (GINSBERG et al. 1994) due to an oligoserine repeat (CAG) that begins around amino acid 300 in the C terminus of the protein. *E2F4* cDNAs were cloned that contained 11, 13 or 16 of these repeats. Analysis of DNA from 55 individuals indicated that all samples contained a common allele of 13 repeats; 7 of the samples also contained an allele with 7, 9, 11, 14, or 16 repeats. Because no functional differences have been ascertained between E2F4 proteins containing 11 or 16 repeats, the significance of the polymorphism is unclear. Although a complete exon/intron structure of the E2F family members has not been reported, a common feature of human and mouse *E2F1* is the presence of a very large first intron located just 5' of the DNA-binding domain (HSIAO et al. 1994; JOHNSON et al. 1994b; LEES et al. 1993; NEUMAN et al. 1994). Genomic DNA containing the transcriptional promotor has been cloned for human and mouse *E2F1*. The *E2F1* promoters from these two species are remarkably similar, having eight mismatches from –57 to +100 (see HSIAO et al. 1994 for a sequence comparison).

1.2 DP Genes

A different approach was used to clone the E2F family member, *DP1* (GIRLING et al. 1993). Investigators studying the *E2* promoter determined that an activity that binds to E2F sites is abundant in F9 embryonal carcinoma cells before, but not after, the cells are induced to terminally differentiate; this activity was named DRTF1 (differentiation-regulated transcription factor 1; LA THANGUE and RIGBY 1987). Protein sequence obtained from purified DRTF1 led to the cloning of a partial cDNA using degenerate polymerase chain reaction primers; a longer cDNA was subsequently isolated from a mouse F9 cDNA library. The cDNA obtained had some homology to human *E2F1* in the sequence corresponding to the DNA-binding domain and the encoded protein could bind to E2F sites. However, the cDNA was distinct from the other E2Fs and was named DP1 (DRTF protein 1). Unlike the results using E2F1-specific antibodies, an antibody specific to DP1 supershifted the majority of the protein species that bind to an E2F site, suggesting that most E2F cellular complexes contain DP1. Using the mouse probe, human DP1 and a highly related protein, DP2, were obtained (HELIN et al. 1993b; WU et al. 1995). Antibodies specific for DP2 supershift the remainder of the E2F activity that is not reactive with the DP1 antibody, suggesting that DP1 and DP2 may be the only DP proteins. No information concerning the gene structure or chromosomal location has been reported for DP1 or DP2.

In summary, the protein sequence of four different mammalian E2Fs and two different mammalian DPs has been reported; at least one other mammalian E2F gene, E2F5, has been cloned. E2F activity has been identified in other species such as *Drosophila* (DYNLACHT et al. 1994a; OHTANI and NEVINS 1994), *S. cerevisiae* (MAI and LIPP 1993), *S. pombe* (MALHOTRA et al. 1993) and *Xenopus* (PHILPOTT and FRIEND 1994). Of these, only the *Drosophila E2F1* and *DP1* genes have been cloned. Details concerning the cloning of the different E2F family members, the sizes of the proteins and mRNAs, as well as the chromosomal location (when known) are reported in Table 1.

Table 1. Mammalian E2F family members

Family member	Source of cDNA	Location	mRNA (kb)	Amino acids	Accession #	Reference
Human *E2F1*	HeLa, Nalm 6[a]	20q11	3.1	437	M96577	(HELIN et al. 1992; KAELIN et al. 1992; SAITO et al. 1995; SHAN et al. 1992)
Human *E2F2*	HeLa, Nalm 6	1p36	6[c]	437	L22846	(IVEY-HOYLE et al. 1993; LEES et al. 1993)
Human *E2F3*	Nalm 6, fetal brain	6p22	6	425	NR	(LEES et al. 1993)
Human *E2F4*	Fetal liver, HeLa, Nalm 6, T84[b]	16q22	2.1–2.9	413[e]	U15641	(SARDET et al. 1995; BEIJERSBERGEN et al. 1994; GINSBERG et al. 1994)
Human *E2F5*	Fetal liver	NR	2.8	345	U15642	(SARDET et al. 1995; R. Bernards, personal communication)
Human *DP1*	Nalm 6	NR	3	410	L23959	(HELIN et al. 1993b)
Human *DP2*	Nalm 6	NR	1.5–10[d]	385	L40386	(WU et al. 1995)
Mouse *E2F1*	3T3	2	2.2,2.7	430	L21973	(LI et al. 1994)
Mouse *E2F2*	NR	NR	NR	NR	NR	(M. Greenberg, personal communication)
Mouse *E2F3*	3T3	NR	NR	NR	NR	(C.J. Fry et al., unpublished data)
Mouse *E2F4*	Mouse embryo	NR	NR	NR	NR	(BEIJERSBERGEN et al. 1994)
Mouse *DP1*	F9 carcinoma	NR	2.8	410	X72310	(GIRLING et al. 1993)

NR, Not reported.
[a] Pre-B cell leukemia line.
[b] Colon carcinoma cell line.
[c] 2.5 kb in skeletal muscle.
[d] Major transcripts are 10 kb in muscle, 1.5 and 2.5 kb in placenta, 1.5 and 2.0 kb in heart, and 2 and 3 kb in liver.
[e] The number of amino acids varies depending on the number of oligoserines encoded at a particular allele.

2 Structure of E2F Proteins

The E2F and DP proteins range from 345 to 437 amino acids (see Table 1) and contain several regions of high sequence similarity that encode functional domains (Fig. 1). The most extensive analysis of the function of different regions of an E2F family member has been performed using human E2F1. Therefore the structure of this protein is described in detail. Initial structural analysis (for review see FARNHAM et al. 1993) of the E2F1 protein defined domains responsible for DNA binding (composed of a basic region and a helix-loop-helix region), transcriptional activation, and Rb binding; further analysis has also identified domains for cyclin A binding, p107 binding, homodimerization, and heterodimerization. These domains in the E2F1 protein are compared, where possible, to the other E2F and DP family members.

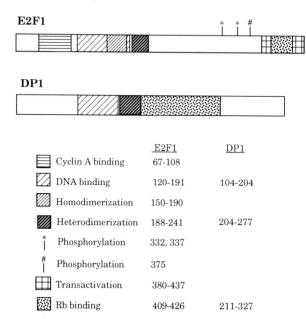

Fig. 1. Schematic representation of the functional domains of the human E2F1 and DP1 proteins. The approximate location (in amino acids number) of each domain discussed in the text is shown

2.1 DNA-Binding Domain

Consistent with the original description of E2F1 as a basic helix-loop-helix (bHLH) protein, CRESS et al. (1993) found that mutations within the two putative helices abolish DNA binding in vitro. However, two studies examining DNA-binding activities of E2F1 mutants have concluded that the E2F1 basic region (originally defined as amino acids 109–128) is dissimilar to that of other bHLH proteins. For example, removal of half of the basic region by truncation of E2F1 at amino acid 117 or deletion of amino acids 113–120 does not abolish DNA binding, although further deletion to amino acid 128 (the 5' edge of the first helix) does eliminate DNA binding (CRESS et al. 1993; JORDAN et al. 1994). The remainder of the basic region, amino acids 117–128, contains only 5 of the 11 amino acids that are normally conserved among bHLH proteins. Amino acid substitutions suggest that these five amino acids are important, although sequence constraints are not as tight as for other bHLH proteins. Although the basic region in many bHLH proteins creates an α-helix that interacts with DNA in the major groove (FISHER et al. 1993), the amino acids in the E2F1-basic region predict a turn. This turn may be critical for binding specificity or affinity since mutations in E2F1 that are predicted to eliminate the turn and/or to create a helix drastically inhibit binding. In summary, mutational analysis suggests that the bHLH region of the E2F1 protein (now defined as amino acids 117–128) is required for DNA binding; however, this region probably has a different structure than many other bHLH proteins. These structural differences may explain why E2F1 binds a different DNA sequence, TTTSSCGC (S=C or G), than many other bHLH proteins (such as the Myc/Max heterodimer) which bind to CACGTG.

Although the other E2F proteins have not been examined in as much detail, the DNA-binding domains have been compared. The DNA-binding domains of E2F1, E2F2, E2F3, and E2F4 are all located near the N terminus of the protein and are 70% identical. The DNA-binding domains of DP1 and DP2 are also located at the N terminus and are approximately 90% identical (Wu et al. 1995). In contrast, the E2F and DP DNA binding domains are only about 40% identical (GIRLING et al. 1993), suggesting that two categories of E2F proteins exist.

2.2 Dimerization Domains

Initial studies demonstrated that bacterially expressed E2F1 could bind to DNA without other protein partners. This DNA binding is dependent upon a homo-dimerization domain which spans amino acids 150–191 (IVEY-HOYLE et al. 1993). Other studies of E2F activity predict heterodimer formation. For example, affinity-purified E2F can be separated into five bands ranging in molecular weight from 50 to 60 kDa by polyacrylamide gel electrophoresis and E2F-binding activity can be reconstituted by mixing a lower and higher molecular weight band (HUBER et al. 1993). Once DP1 was cloned, it became apparent that this is a major component of the E2F activity, and several groups quickly demonstrated that E2F1 and DP1 can form heterodimers using bacterially expressed protein (KREK et al. 1993), a yeast two hybrid system (BANDARA et al. 1993), glutathione S-transferase fusion columns (HELIN et al. 1993b), or immunoprecipitation of transfected proteins from mammalian cell extracts (HELIN et al. 1993b). The heterodimers of E2F1 and DP1 are more active in DNA binding and transactivation than are the homodimeric forms of either protein (BANDARA et al. 1993; HELIN et al. 1993b; KREK et al. 1993). The region of E2F1 required for heterodimerization includes the leucine-zipper region (amino acids 191–284) found adjacent to the DNA-binding domain (HELIN et al. 1993b; JORDAN et al. 1994; KREK et al. 1993). The region of DP1 required for heterodimerization (amino acids 205–277) is also adjacent to the DNA-binding domain (HELIN et al. 1993b).

2.3 Rb- and p107-Binding Domains

Both Rb and p107 can bind to the carboxy terminus of E2F1 in vitro (amino acids 409–426). Although initial experiments suggested that these domains were identical, further analysis has identified specific amino acids that are more critical for binding of E2F1 to p107 than to Rb. Mutation of amino acids 411, 413, or 421 in E2F1 reduces binding of p107 without affecting E2F1/Rb interactions (CRESS et al. 1993). E2F1 proteins containing a combination of the 411 and 421 muta-tion, or a deletion of amino acids 420–422, no longer respond to p107 in a transfection assay. The transcriptional activity of the 411/421-E2F1 mutant construct is equal to that of wild-type E2F1 in a cotransfection assay with the dihydrofolate reductase (dhfr) promoter, suggesting that p107 binding is not critical for E2F1-mediated transcription. This hypothesis is supported by recent studies demonstrating that in cells E2F1 binds preferentially to Rb, not to p107. A separate study found that mutation of amino acid 411 eliminates Rb binding (HELIN et al. 1993). Perhaps the exact amino acid substituted at position 411 is critical in determining E2F1/Rb interactions. Bandara demonstrated that a protein domain in the C-terminus of mouse DP1 (amino acids 211 to 327) can bind weakly to Rb, and that this interaction stabilizes the binding of the E2F1/DP1 heterodimer to Rb (BANDARA et al. 1994).

E2F4 associates preferentially with p107, and not with Rb, in cells. Although the precise amino acids responsible for this interaction are not known, one of the E2F4 cDNAs obtained using p107 as a probe encoded only the last 31 amino acids. This region corresponds to the region of E2F1 which is required for interaction with Rb and p107. Transfection studies suggest that dimerization of E2F4 with DP1 is required not only for efficient DNA binding but also for interaction with p107 (BEIJERSBERGEN et al. 1994), suggesting that domains of both E2F4 and DP1 contribute to stable binding to p107.

2.4 Transactivation Domain

Fusion of portions of the C-terminus of the E2F1 protein to a GAL4 DNA-binding domain allowed an initial demonstration that amino acids 368–437 could function as a transactivation domain (KAELIN et al. 1992). Deletion of the extreme C-terminus (amino acids 417–437 or 409–437) abolishes the transcriptional activity of a GAL4/E2F1 fusion protein in the Burkitt's lymphoma cell line DG75 and of an E2F1 protein in T98G glioblastoma cells (CRESS et al. 1993; KAELIN et al. 1992). These studies suggested that the minimal Rb-binding domain (located from 409–426) was important for transcriptional activity of E2F1. However, further analysis indicated that the Rb-binding domain does not have independent transcriptional activity. Two copies of a repeated motif (E/DFXXLXP) are located just upstream and one copy just downstream of the Rb-binding domain. Deletion analysis indicates that E2F1-protein sequences containing the upstream or downstream repeat elements contribute to transcriptional activity in SAOS-2 cells by synergizing with some component of the Rb-binding domain (HAGEMEIER et al. 1993). In summary, these studies indicate that amino acids 380–437 comprise a potent transactivation domain with elements upstream of, downstream of, and including the minimal Rb-binding domain. In contrast, other studies have shown that deletion of the C-terminal 417–437 amino acids (which includes the Rb-binding domain and the downstream repeat element) has little effect on the transcriptional activity of a GAL4/E2F1 fusion protein in a variety of cell types, including SAOS-2 cells (HELIN et al. 1993a; SHAN et al. 1992). The causes of discrepancy in these reports are unknown; further analysis is required to determine whether the transcriptional activity of E2F1 can indeed be dissociated completely from the Rb-binding domain.

2.5 Cyclin-Binding Domains

E2F family members can also associate with cyclin E and cyclin A, and the abundance of these E2F/cyclin complexes changes in different stages of the cell cycle (LEES et al. 1992; see also COBRINIK, this volume). Recent studies suggest that the E2F1/DP1 heterodimer is phosphorylated at distinct times in the cell cycle due to different cyclin/kinase interactions. First, a cyclin-dependent phosphorylation in late G_1 of two serines on E2F1, amino acids 332 and 337, can inhibit Rb-E2F1 interactions (FAGAN et al. 1994). The timing of expression of this phosphorylation event suggests that in cells it may be mediated by cyclin E. The next wave of E2F phosphorylation is likely due to cyclin A/cdk2. E2F1 (and possibly E2F2 and E2F3 since they contain similar amino acid sequences) can bind directly to cyclin A via amino acids 67–108 (KREK et al. 1994). In vitro an E2F1/DP1 heterodimer can be phosphorylated by cyclin A/cdk2, resulting in a protein complex with reduced DNA-binding affinity (DYNLACHT et al. 1994b). Preliminary studies indicate that DP1 is the target of the cyclin A-dependent kinase; prior treatment of DP1 with the kinase can inhibit DNA binding (DYNLACHT et al. 1994b) and DP1 is phosphorylated maximally in late S phase, at about the peak of cyclin A expression (BANDARA et al. 1994). In vivo studies using an E2F1 mutant lacking the cyclin A binding domain also suggest that the phosphorylation of DP1, but not E2F1, is dependent on cyclin A (KREK et al. 1994). In contrast, others suggest that cyclin A/cdk2-mediated phosphorylation of E2F1 inhibits binding of heterodimeric E2F1/DP1 to DNA (XU et al. 1994). Perhaps phosphorylation of either E2F1 or DP1 by cyclin A contributes to loss of E2F DNA-binding activity in S phase. It is not known which amino acids of E2F1 or DP1 are phosphorylated in the inactive form of these proteins. A third wave of phosphorylation of E2F may come in the G_2/M phase due to a peak of activity of cyclin A complexes with cdc2. PEEPER et al. (1995) have shown that the cyclin A/cdc2 complex can efficiently phosphorylate E2F1 on serine 375 when coexpressed in insect cells. This amino acid is normally phosphorylated in human cells, but the amount of phosphorylation on this site in different stages of the cell cycle has not yet been determined. They have also shown that Rb has a five- to tenfold higher affinity for E2F1 molecules that are phosphorylated on this specific amino acid. Serine 375 is located in a leucine-serine-proline triplet that is conserved in E2F1, E2F2, and E2F3, but not in E2F4.

Thus, a model for cell cycle-dependent activation and inactivation of E2F by phosphorylation can be proposed: in late G_1 cells E2F1 is phosphorylated on amino acids 332 and 337 by cyclin E/cdk2 and is released from Rb as an active transcription factor; in middle to late S phase E2F1 and/or DP1 is phosphorylated by cyclin A/cdk2, causing disruption of the heterodimer and loss of DNA-binding activity; in G_2/M phase E2F1 is then phosphorylated on amino acid 375 to enable high-affinity binding by Rb in the subsequent G_1 phase.

In summary, functional protein domains for E2F1 include the bHLH region, dimerization region, transactivation domain, pocket protein (e.g., Rb, p107, or p130) binding domain, cyclin A binding domain, and cdk-targeted serines (Fig. 2). The other E2Fs have about 70% sequence identity to E2F1 in their DNA-binding

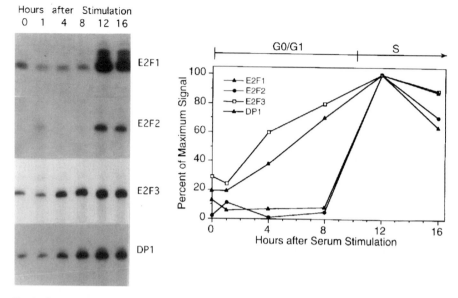

Fig. 2. Growth regulation of E2F family members. Serum-starved NIH 3T3 cells were serum stimulated for the indicated times. Cytoplasmic RNA was prepared for RNAse protection analysis and then hybridized with mouse E2F1, E2F2, E2F3, and DP1 probes (as described in SLANSKY et al. 1993). For analysis of E2F1 and DP1, 10 µg of RNA was analyzed per time point; an overnight exposure of the gel is shown. For E2F2, 60 µg of RNA was analyzed per time point; a 3-day exposure of the gel is shown. For E2F3, 40 µg of RNA was analyzed per time point; an overnight exposure of the gel is shown. *Left,* results of these RNAse protection experiments; *right,* graphic representation from the same experiments. The signals were quantitated on a Phosphorimager and plotted relative to the maximum value (the 12-h samples) for each individual mRNA

domains and can each heterodimerize with either DP1 or DP2 (Wu et al. 1995). The transactivation domain is also highly conserved in all E2Fs, although this region of the different family members is thought to confer in vivo specificity of binding to Rb (E2F1, E2F2, E2F3) or p107 (E2F4, E2F5). E2F1, E2F2, and E2F3 are also predicted to bind directly to cyclin A, whereas E2F4 and E2F5 probably require binding to p107 or to p130 to mediate a cyclin A/cdk2 interaction. The DP proteins are similar to the E2Fs only in the DNA-binding and heterodimerization domains and are not thought to contain a transactivation domain, to bind to cyclins, or to specify Rb versus p107 interactions.

3 Expression Patterns of E2F Family Members

An important question concerning the E2F family is whether individual family members have distinct or redundant roles. This section addresses possible tissue-, developmental-, and/or temporal-specific expression of the different E2F and DP genes. Sections 4 and 5 discuss the possibility that specificity is achieved by differential activity of the E2F family members.

3.1 Tissue-Specific Expression

The expression of the different E2F family members has been examined in Northern analysis using RNA from adult human tissues (BEIJERSBERGEN et al. 1994; GINSBERG et al. 1994; HELIN et al. 1992; IVEY-HOYLE et al. 1993; LEES et al. 1993; WU et al. 1995). The approximately 3.1-kb *E2F1* mRNA is most easily detected in the heart, brain, placenta and lung; lower levels of *E2F1* can be detected in the liver, skeletal muscle, kidney and pancreas. The approximately 6-kb *E2F2* mRNA is most readily detectable in the placenta, followed by lung and kidney. All of the tissues analyzed express a 2.2-kb *E2F4* mRNA and a 6-kb *E2F3* mRNA, except the brain which does not express detectable levels of *E2F3* (BEIJERSBERGEN et al. 1994; GINSBERG et al. 1994). The approximately 3-kb *DP1* mRNA is strongly detected in the muscle, placenta, liver, kidney, and brain but is not detectable in mRNA from the heart. *DP2* mRNA, on the other hand, is expressed in most tissues, including the heart, but not in the brain or lung. These analyses suggest that heterodimers of E2F1 and E2F4 with DP1 may be important mediators of E2F-regulated transcription in the brain. In contrast, DP2 may be the critical DP partner in the heart. However, since most of the family members are expressed in a wide variety of tissues (summarized in Table 2), it is unlikely that distinct roles of the individual E2Fs derive solely from tissue specificity.

3.2 Developmental Regulation

Examination of the relative abundance of the different family members in adult tissues does not address the possibility of differential expression of the transcription factors during development. Such experiments are being performed in a mouse model system, where development can be easily monitored.

Table 2. Expression of E2F family members in adult human tissues

E2F	Heart	Brain	Placenta	Lung	Liver	Muscle	Kidney	Pancreas	Reference
E2F1	++	++	++	++	+	+	+	+	(HELIN et al. 1992)
E2F2	–	–	++	+	–	–	+	–	(IVEY-HOYLE et al. 1993)
E2F3	+	–	++	++	++	++	++	+	(LEES et al. 1993)
E2F4	*	*	*	*	*	*	*	*	(GINSBERG et al. 1994)
DP1	–	++	++	+	++	++	++	+	(WU et al. 1995)
DP2	+	–	++	–	++	++	++	+	(WU et al. 1995)

The relative expression level of each family member in the different tissues is indicated; high (++), low (+), or not detectable (–). *Asterisks*, E2F4 was detected, but relative expression levels were not reported. Expression levels can be compared only within each family member due to variation in specific activity of the probes, exposure times, and hybridization conditions.

Preliminary in situ hybridization experiments (G. LYONS, personal communication) using an *E2F1* cDNA probe with a developmental series of C57BL/6 and BALB/c mouse embryos showed that *E2F1* mRNA is initially widespread at 8.5 days postcoitum. By 11.5 days postcoitum the *E2F1* expression pattern becomes restricted; the transcripts are most abundant in the ependymal layer of the neural tube and the subventricular zone of the developing brain (areas which contain rapidly dividing neurons). There is lower signal detected in the liver, limb bud mesenchyme, nasal sinus epithelium, and heart. At 13.5 days postcoitum the highest level of *E2F1* mRNA is in the liver and the ependymal and subventricular layers of the central nervous system; lower signals can be detected in the pituitary gland, lung, kidney, nasal sinus epithelium, gut, and peridigital mesenchyme of the limb. At birth *E2F1* mRNA is detected in the external granule layer of the cerebellum, and in the adult E2F1 *mRNA* is found in the granule and molecular layers of the cerebellum and at low levels in the hippocampus and cortex in the mouse. Studies of the expression of the other E2F family members during mouse development are in progress.

A different developmental study of *E2F1* and *DP1* in the mouse began with embryonic day 14 and continued through the postnatal period and into adult (TEVOSIAN et al. 1995). Analysis of total tissue RNA demonstrated that expression of *E2F1* and *DP1* mRNA is high in the period from embryonic day 18 to postnatal day 3; expression declines following birth with minimal expression in the adult. High-resolution in situ analysis indicated that there is cell- and region-specific expression of *E2F1* and *DP1*, and that the expression of E2F1 is not completely correlated with cells of highest proliferative potential. For example, in the liver E2F1 expression is high in dividing hepatocytes but not in proliferating hematopoietic islands.

3.3 Cell Cycle Expression

Expression of the different E2F family members during specific stages of the cell cycle can also provide functional selectivity. To address this possibility the mRNA levels, promoter activity, and protein levels of the different family members must be examined. A complete analysis of this type has been performed for E2F1. The levels of *E2F1* mRNA increase at the G_1/S phase boundary in serum-stimulated NIH 3T3 cells (SLANSKY et al. 1993), mitogen-activated T cells (KAELIN et al. 1992), regenerating liver (BENNETT et al. 1995), EGF-stimulated keratinocytes (SATTERWHITE et al. 1994), and mitogen-stimulated human umbilical vein endothelial cells (ZHOU et al. 1994). The levels of *E2F1* promoter activity increase at the G_1/S phase boundary in cells stimulated from quiesence (HSIAO et al. 1994; JOHNSON et al. 1994b; NEUMAN et al. 1994) and in proliferating cells synchronized by mitotic selection (K.E. Boyd and P.J. Farnham, unpublished data). Also, the amount of E2F1 protein that can be measured by gel mobility shift assays peaks in S phase (CHITTENDEN et al. 1993).

Initial investigations into the regulation of other E2F family members have shown that the levels of *E2F3, E2F4,* and *DP1* mRNAs in mitogen-stimulated kerotinocytes and NIH 3T3 cells do not change dramatically through the cell cycle (GINSBERG et al. 1994; LI et al. 1994; SATTERWHITE et al. 1994; SARDET et al. 1995). Also, E2F4 protein levels are constitutive after serum stimulation of NIH 3T3 cells (GINSBERG et al. 1994). Levels of E2F5 increase 12-fold in middle G_1 after serum stimulation of kerotinocytes (SARDET et al. 1995). The levels of *E2F2* are very low and are difficult to analyze; however, it has been reported that levels of *E2F2* mRNA do not significantly change after EGF stimulation of mink lung epithelial cells (SATTERWHITE et al. 1994). The response of *E2F2* to stimulation with the more complete set of growth factors found in serum has not been reported.

Since many of the studies of the regulation of E2F family members have used Northern analysis, a quantitative measure of the levels of the E2F mRNAs in different stages of the cell cycle was not possible due to the low sensitivity of this assay. We have therefore examined the steady-state levels of RNA from different E2F family members throughout the growth cycle of NIH 3T3 cells using quantitative RNAse protection assays. The cells were serum-starved for 44 h to induce a quiescent state. Then, serum-containing medium was added, and cells were harvested at different time points to examine the levels of E2F family members in different phases of the growth cycle. As shown in Fig. 2, the expression of E2F mRNAs falls into two categories. *E2F1* and *E2F2* both peak in S phase, whereas the levels of *E2F3* and *DP1* rise early in G_1 and do not increase much more as cells enter S phase. These results are similar to Northern analyses performed using RNA after stimulation of human T cells (J. Lees, unpublished data). *E2F1* and *E2F2* are both expressed in a cell cycle dependent manner, with *E2F2* being expressed after *E2F1*. In contrast, both *E2F3* and *E2F4* are present in quiescent cells with *E2F4* being 20-fold more abundant than *E2F3*; the levels of *E2F3* and *E2F4* increase slightly after stimulation of T cells.

In summary, the presence of more than one E2F in any particular tissue and the constitutive expression of several of the E2F family members through the cell cycle suggest that if specificity occurs, it must be attained via distinct features (perhaps specified by unique protein domains) of the different heterodimeric complexes.

4 Regulation of Cellular Promoters by E2F

Although several viruses have evolved efficient strategies to increase E2F activity, only adenovirus contains E2F-regulated promoters, suggesting that key targets of E2F transcription factors may be cellular promoters. Cellular targets of E2F were first identified through mutational analysis of cell cycle regulated promoters. In 1989 it was discovered that E2F sites are important for the activity of two cellular promoters, *dhfr* and *c-myc* (BLAKE and AZIZKHAN 1989; HIEBERT et al. 1989; THALMEIER et al. 1989). Subsequently, the E2F sites in these promoters were also shown to be required for activation by viral oncogenes and for maximal activity in response to serum stimulation (HIEBERT et al. 1991; MEANS et al. 1992; MOBERG et al. 1992; SLANSKY and FARNHAM 1993; THALMEIER et al. 1989; WADE et al. 1992). Additional cellular and viral promoters that contain E2F sites have been identified by sequence inspection and mutational analysis; a detailed description of these E2F sites can be found in Table 3. Cellular promoters whose E2F sites contribute to transcriptional regulation include genes required for DNA synthesis (*dhfr*, thymidine kinase, DNA polymerase-α), transcriptional regulators of cell growth (N-*myc*, c-*myc*, retinoblastoma, *cdc2*, *E2F1*, and b-*myb*), and growth factors (*IGF-1*).

4.1 Requirements for Growth Regulation

The expression of several of the genes listed in Table 3 is cell cycle regulated. In many cases the contribution of the E2F site to regulation is inferred from effects seen by large scale deletions of promoter DNA, not by mutation of individual E2F binding sites. However, a transient transfection system has been used to demonstrate that an E2F site is required for growth regulation of the *dhfr*, b-*myb*, and *E2F1* promoters. In these experiments the cells are forced into quiescence, then stimulated to enter the growth cycle synchronously. By analysis of promoter activity at various points after stimulation the growth regulation of a wild-type or mutated promoter can be determined.

The *dhfr* promoter contains two inverted, overlapping E2F sites at the transcription initiation site. Transcription from a *dhfr* promoter fragment increases approximately 15-fold at the G_1/S phase transition; mutation of the E2F sites results in a promoter whose activity increases only threefold at the G_1/S phase transition (MEANS et al. 1992). The b-*myb* promoter contains E2F sites located at –200 (relative to the translation initiation site). Transcription from a b-*myb* promoter fragment increases tenfold at the G_1/S phase boundary; activity from a b-*myb* promoter with mutated E2F sites does not significantly change during the growth cycle (LAM and WATSON 1993). The *E2F1* promoter contains two sets of overlapping, inverted E2F sites spanning –10 to –40 (relative to the transcription initation site). Transcription from the mouse E2F1 promoter was reduced from 42-fold to fivefold when the E2F sites were mutated (HSIAO et al. 1994); similar results were obtained with the human promoter (JOHNSON et al. 1994b; NEUMAN et al. 1994).

Somewhat controversial results have been obtained regarding the role of the E2F sites in the c-*myc* promoter. Although one study demonstrated that a three fold increase in c-*myc* promoter activity seen 4 h after serum stimulation of resting cells is dependent on the E2F site (MUDRYJ et al. 1990), others have failed to see growth regulation of the c-*myc* promoter or activation of this promoter by E2F1 (LI et al. 1994; J.E. Slansky and P.J. Farnham, unpublished data). However, the E2F sites in the c-*myc* promoter have been more clearly implicated in regulation of promoter activity in differentiating cells (MELAMED et al. 1993) and in transactivation by viral oncoproteins (HIEBERT et al. 1989; MOBERG et al. 1992; THALMEIER et al. 1989).

It was initially proposed that increased binding of E2F to cellular promoters is sufficient for increased expression at the G_1/S phase boundary. In support of this hypothesis two copies of a 29-bp oligonucleotide containing the *dhfr* E2F sites are sufficient for G_1/S phase-specific transcription (SLANSKY et al. 1993). However, accumulating evidence suggests that the presence of an E2F site may not be sufficient for G_1/S phase activation. For example, deletion of sequences upstream of the E2F sites in the *dhfr* (SLANSKY and FARNHAM 1993), *E2F1* (HSIAO et al. 1994; JOHNSON et al. 1994b; NEUMAN et al. 1994), b-*myb* (LAM and WATSON 1993), and DNA polymerase-α (PEARSON et al. 1991) promoters greatly reduces regulated activity. Additional evidence which supports the hypothesis that E2F sites are not sufficient for regulated expression comes from cellular promoters that are not growth regulated. Deoxycytidine kinase has an E2F site in its promoter, but does not display cell cycle-dependent regulation (HENGSTSCHLÄGER et al. 1994). Also, a consensus E2F site can be found in the *Rb* promoter even though activity from this promoter changes only twofold during the growth cycle (HENGSTSCHLÄGER et al. 1994; SHAN et al. 1994; J.E. Slansky and P.J. Farnham, unpublished data) and mutation of the E2F site has very little affect on promoter activity in growing cells (GILL et al. 1994). Although it is not clear why two copies of the *dhfr* E2F sites can confer regulation, perhaps the particular orientation and spacing of the sites in this construct provide a unique promoter structure. Another possibility is suggested by the observation that in some cells protein binding can be detected to a site immediately adjacent to the E2F sites (J. CAMPISI, personal communication); this region is present in the 20-bp oligonucleotide used in the synthetic constructs.

One hypothesis to explain why E2F sites are necessary, but not sufficient, for growth regulation is that other transcription factors work in concert with E2F to specify growth-regulated transcription. Constructs in which other transcription factor binding sites were placed upstream of the E2F sites from the *dhfr* (J.E. Slansky and P.J. Farnham, unpublished data) or *E2F1* (K.M. Hsiao and P.J. Farnham, unpublished data) promoter display a larger increase in promoter activity at the G_1/S phase boundary than do similar constructs containing the E2F sites alone. In particular, synthetic oligonucleotides containing binding sites for Sp1 and CCAAT factors can stimulate growth regulation mediated by E2F sites. Sp1 has been previously implicated in growth regulation; deletion of Sp1 binding sites in certain cellular promoters reduces growth-regulated transcription (MILTENBERGER et al. 1995; SCHILLING and FARNHAM 1995; SLANSKY and FARNHAM 1993), and Sp1 protein can interact with the Rb protein (CHITTENDEN et al. 1991a; KIM et al. 1992; UDVADIA et al. 1993). CCAAT elements contribute to the activation of the *E2F1* promoter in S phase (HSIAO et al. 1994) and binding of proteins to CCAAT elements has been implicated in the regulation of other serum-responsive promoters (CHANG and CHENG 1993; DUTTA et al. 1990; MARTINELLI and HEINTZ 1994; PANG and CHEN 1993). Perhaps E2F family members contain domains that interact with Sp1 or CCAAT factors to stabilize binding to the promoter. Alternatively, a synergy between the transactivation domain of E2F family members and

Table 3. Potential targets for regulation by E2F

Species/gene	E2F site	Ori.	Bind.	Reg.	Sp1	CAT	Reference
Human cdc2	−132 tttcTTTCGCGCtcta	↑	C	R	√	√	(DALTON 1992)[c,d] R^b / (FURUKAWA et al. 1994)[c] R^a C
Human deoxycytosine kinase	−7 tgacTTTGGCGCgcgg	↓	?	N	√	–	(SONG et al. 1993)[c] / (HENGSTSCHLÄGER et al. 1994) N^b
Hamster dhfr	−3 gcaaTTTCGCGCCAAActtg	↑	C / P	R / V / e	√	√	(MARIANI et al. 1981) R^b / (MITCHELL et al. 1986)[c] R^a / (BLAKE and AZIZKHAN 1989) Ce / (HIEBERT et al. 1991) CPV
Human dhfr	+1 acaaTTTCGCGCCAAActtg	↑	C	R	√	–	(SHIMADA et al. 1986) C / (GOLDSMITH et al. 1986) R^a / (HENGSTSCHLÄGER et al. 1994) R^b
Mouse dhfr	−319 cggaTTTCCCGCgggg	↑	?	R / E	√	√	(SANTIAGO et al. 1984) R^a / (FARNHAM and SCHIMKE 1986) R^b
	−14 gcgaTTTCGCGCCAAActtg	↑	C / P / B	V			(MEANS et al. 1992) R^a CP / (SLANSKY et al. 1993) R^a TV / (Li et al. 1994) CBT
Human DNA polymerase-α	−124 cgcgTTTGGCGCcctg	↓	C	R	√	√	(PEARSON et al. 1991)[c] R^a
	−107 cggcCTTCCCGCggac	↓					(Li et al. 1994) T
Adenovirus E2	−72 tagtTTTCGCGCttaa	↑	C / P / B	V	–	–	(KOVESDI et al. 1986) C / (NEVINS 1992) review V
	−33 ctagTTTCGCGCcctt	↓					(Cress and Nevins, this volume)
Human E2F1	−35 gctcTTTCGCGGCAAAagg	↑	C / P	R / E / V	√	√	(KAELIN et al. 1992) R^a / (JOHNSON et al. 1994b)[c,d] R^a CPVT / (NEUMAN et al. 1994)[c] R^a C
	−18 aggaTTTGGCGCGTAAaagt	↑					
Mouse E2F1	−44 gctcTTTCGCGGCAAAagg	↑	C	R / E	√	√	(SLANSKY et al. 1993) R^a / (HSIAO et al. 1994)[c] R^a CT
	−27 aggaTTTGGCGCGTAAaagt	↑					
Rat insulin-like growth factor1	−41 ttttTTTCCCCGCcctt	↓	C	R / V	√	?	(PORCU et al. 1992) R^a V / (PORCU et al. 1994) CVT
Mouse b-myb	−197 [atg]cctaTCTCCCGCCAAGtgcg	↓	C	R / E / V	√	–	(LAM and WATSON 1993)[c] R^a VC / (LAM et al. 1994) CV
Mouse c-myb	−282 [atg]cagaTTTGGCGGgagg	↑	C		√	–	(WATSON et al. 1987)[c] / (MUDRYJ et al. 1990) C

Gene	Position	Sequence	Ori.	Bind	Reg	Check	Reference
Human c-myc	-53	P2tcttTTCCCGCCAAGcctc	↓	C / B	V / e	- / √	(HIEBERT et al. 1989) CV (MUDRYJ et al. 1990) Ce (THALMEIER et al. 1989)d CV (OSWALD et al. 1994) BT (LI et al. 1994) T
Mouse c-myc	-59	P2tcttTTCCCGCCAAGcgtc	↓	C	√	?	(MOBERG et al. 1992) C (ROUSSEL et al. 1994) T
Human n-myc	-188	ggctTTGGCGCGAAAgcct	↑	C	√	-	(TAKEHANA et al. 1991)c (HARA et al. 1993) C
Mouse n-myc	-146	ggctTTGGCGCGAAAggct	↑	C	V	√	(DEPINHO et al. 1986)c (MUDRYJ et al. 1990) C (HIEBERT et al. 1991) CV
	-131	aggcTTGGCGCctcc	↑			-	
Human Rb	-190	acgtTTCCCGCggtt	↑	C / P	N	?	(SHAN et al. 1994) NT (PARK et al. 1994) P (LI et al. 1994) T
Human thymidine kinase	-97	gagaTTGGCCGcagc	↓	C / E	R	√	(KIM and LEE 1991)c Ra (KIM and LEE 1992) CRa (LI et al. 1993) EC (LI et al. 1994) T (HENGSTSCHLÄGER et al. 1994) Rb
	-104	caaaTCTCCCGCagg	↑				
Mouse thymidine kinase	-85	ttgaGTTCGCGGGCAAAtgcg	↑	C / B	R / V	√	(DOU et al. 1994) BC (OGRIS et al. 1993) Ra CV

Second column, the E2F sites contained within the 5' flanking sequences of the gene. Number to the left of the sequence, the position of the first nucleotide relative to the transcription initiation site unless it is followed by atg or p2, then the numbers are relative to the translation start site or the P2 promoter, respectively. Upper-case letters, the sequence included in the consensus E2F binding site (TTTSSCGC) in the 5' to 3' orientation (in several promoters there may be two overlapping, inverted consensus sites); lower-case letters, sequences flanking the E2F site. Orientation (Ori.) of the site with respect to the coding sequence is indicated by the direction of the arrow: → the site is on the coding strand; ←, the site is on the noncoding strand. Binding (Bind.): **C**, presence of E2F-binding activity in crude extracts; **P**, DNA-binding activity using partially purified E2F; **B**, binding activity using a bacterially expressed E2F family member. The same symbols appear next to the references in which they were reported. Regulation (Reg.), the types examined for that gene: **V**, by viral oncoproteins; **E**, through the E2F sites at the G₁/S boundary; **e**, through E2F sites in growing cells or before the G₁/S boundary (i.e. c-myc); **R**, at the G₁/S boundary that has not yet been examined using mutated E2F sites; **N**, no change in RNA levels was observed at the G₁/S boundary. Check marks, the presence of Sp1 binding sites (GGGCGG) or CCAAT-binding sites (CCAAT) found within 300 bases from the transcription start site. The occasional consensus Sp1 site harbored within an E2F site was not considered; one study has shown that Sp1 does not bind to the E2F site of the human c-myc promoter (THALMEIER et al. 1989). Question marks, the information is not found in the literature. **T**, E2F family member(s) was used in transfection assays with the promoter reporter construct.

a Growth cycle regulation.   b Cell cycle regulation.   c The reference in which the promoter sequence was reported.
d The reference from which promoter coordinates come, in instances of different numbering systems being used for a given promoter.

other transcription factors may be required for maximal transcriptional activation. Evidence that Sp1 can recruit a CCAAT factor to the rat *CYP2D5* promoter suggests the possibility of additional cooperation between Sp1 and CCAAT factors (LEE et al. 1994). It is of note that all of the cellular promoters described in Table 3 contain either an Sp1 or CCAAT element. Further mutational analysis of these promoters is essential to determine whether these different transactivators cooperate with E2F family members to mediate growth-regulated transcription.

4.2 Activation Versus Repression

Mutational analysis of some promoters suggests that E2F family members are positive activators of transcription. For example, deletion of the E2F sites in the c-*myc* and the *cdc2* promoters results in a 50%–90% loss of activity (FURUKAWA et al. 1994; MOBERG et al. 1992; THALMEIER et al. 1989), mutation of the E2F site in the n-*myc* promoter causes a 90% reduction of activity in undifferentiated NEC14 cells (HARA et al. 1993), and mutation of the E2F site in the hamster *dhfr* promoter results in a 80% reduction of activity in HeLa cells (BLAKE and AZIZKHAN 1989). Also, the E2F sites in numerous promoters are required for regulation by viral transactivators (Table 3). In support of the hypothesis that E2F family members are positive mediators of transcriptional activity, the C-terminal region of E2F1 can function as a transactivator when fused to a heterologous DNA-binding domain (FLEMINGTON et al. 1993; KAELIN et al. 1992). In fact, each of the E2F family members can elicit an increase in activity from promoters containing E2F sites in transient transfection assays (BEIJERSBERGEN et al. 1994; GINSBERG et al. 1994; LI et al. 1994; WU et al. 1995). Also, E2F1 can mediate transcriptional activation in an in vitro assay (DYNLACHT et al. 1994b).

However, other results indicate that E2F family members may also be involved in repression of transcription. For example, mutation of the E2F sites in the mouse *dhfr* promoter and the b-*myb*, n-*myc*, and *E2F1* promoters increases transcriptional activity three- to tenfold (HIEBERT et al. 1991; HSIAO et al. 1994; JOHNSON et al. 1994b; LAM et al. 1994; LAM and WATSON 1993; MEANS et al. 1992; J.E. Slansky and P.J. Farnham, unpublished data). The *IGF-1* promoter is normally not active in quiescent cells; however, mutation of the E2F sites in this promoter allows high level expression in quiescent cells (PORCU et al. 1994). Similarly, addition of an E2F site to an Sp1-driven synthetic promoter, an ATF-driven promoter, or to the *SV40* early promoter reduces transcriptional activity (SLANSKY et al. 1993; WEINTRAUB et al. 1992). Because silencer domains have not been identified in E2F proteins, it is believed that the repression is mediated by protein–protein interactions between different E2F family members and other cellular proteins such as Rb or p107. Both Rb and p107 can bind to the carboxy terminus of E2F proteins. Transfection experiments demonstrate that E2F-mediated transcription can be inhibited by increased levels of either p107 or pRb (DYSON et al. 1993; HIEBERT et al. 1992; SCHWARZ et al. 1993; ZHU et al. 1993), and recent in vitro results have shown that the addition of Rb to a cell-free transcription system activated by E2F1 can directly reduce promoter activity (DYNLACHT et al. 1994b).

Several possible mechanisms, which are not mutually exclusive, have been put forth to explain how Rb and p107 can repress E2F-mediated transcription. It has been proposed that Rb contains a repressor domain which can reduce the activity of a transcription complex (WEINTRAUB et al. 1992). Since the association of Rb with E2F increases the half-life of the E2F/DNA interaction by more than tenfold (HUBER et al. 1994), the affinity of the repressive E2F/Rb complex for DNA is much higher than is the affinity of free E2F. This suggests that changing the balance between binding of the repressive E2F/Rb complex versus free E2F requires more than simply increasing the amount of E2F. The E2F/Rb protein complex differs from free E2F1 not only in DNA binding affinity, but also in the ability to influence DNA structure. Upon binding to DNA, E2F induces a bend (flexure angle) of 125°; in contrast, Rb/E2F complexes induce a bend of only 80°. The effect of p107 on the bending of DNA by E2F has not yet been analyzed. Although the precise consequences of this change in flexure angle are not known, one possible result is the creation of a suboptimal spacing between two proteins in the transcription complex, eliminating critical protein–protein contacts. Another mechanism by which Rb may repress transcription is by blocking the E2F-activation domain from interacting with other transcription factors (discussed in Sect. 4.1) or with other components of the basal transcriptional machinery. E2F1 can bind to the general transcription factor TBP via the same region that is required for binding to Rb (HAGEMEIER et al. 1993), suggesting that Rb may block E2F1/TBP interactions.

A model that incorporates the influences of Rb and cyclin A on E2F1-mediated transcription has been suggested. First, E2F1 must be dissociated from the transcriptional repressing effects of Rb. Using an in vitro transcription system, DYNLACHT et al. (1994b) have demonstrated that cyclin E/cdk2 (whose activity peaks in the late G_1 phase of the cell cycle) can phosphorylate Rb, releasing free E2F. Others have shown that cyclin-dependent phosphorylation in late G_1 of E2F1 can also inhibit Rb-E2F1 interactions (FAGAN et al. 1994). Thus, changes in the phosphorylation status of both E2F1 and Rb in late G_1 can cause a release of E2F1 from Rb, allowing E2F to transactivate cellular promoters. A cyclin A-dependent phosphorylation of the E2F1/DP1 can dissociate the heterodimer and inhibit E2F DNA-binding activity. This model suggests that E2F1 functions as a transcriptional activator only after cyclin E levels increase and before cyclin A levels increase (i.e., late G_1 to middle S phase). This fits well with the observations of cell cycle-dependent E2F-mediated transcription of genes such as *dhfr* and b-*myb*. The above model does not take into account the role of p107 in E2F-mediated transcription; this is the focus of ongoing analyses.

In summary, the role of the E2F site varies in different promoters. However, a common theme is emerging in that the consequences of E2F binding to a promoter appear to be highly influenced by other proteins, such as activators (e.g., CCAAT factors), repressors (e.g., Rb), and basal transcription factors (e.g., TBP). Understanding the exact mechanisms by which E2F family members activate transcription will require the use of a highly defined in vitro system. Unfortunately, most reconstituted transcription systems are extremely inefficient in transcription of E2F-regulated cellular promoters, such as *dhfr*.

5 Can Specific E2F Family Members Be Linked to Specific Target Genes?

The distinction as to whether a particular E2F site functions in specific stages of the cell cycle or mediates activation versus repression may be a result of the binding of specific E2F family members. For example, a particular E2F family member may recognize a specific flanking sequence or structure in addition to the consensus E2F site. Also, the spacing between E2F sites, such as those in the E2 promoter (HARDY and SHENK 1989) may determine optimal binding by the different E2F family members. Alternatively, the presence of binding sites for other transcription factors may determine whether a given E2F can interact stably at specific promoters. This section reviews experimental evidence that may allow us to link specific E2F family members with the regulation of specific genes.

5.1 Can Specificity Be Achieved by DNA-Binding Requirements?

To date all of the family members, when expressed in bacteria, can bind to an oligonucleotide containing a consensus E2F site (5' atttaagTTTCGCGCcctttccaa). However, the affinity of E2F3 for this DNA is slightly less than is the affinity of E2F1 and E2F2 (LEES et al. 1993). This difference in affinity may indicate a true preference of specific E2Fs for this site; alternatively, it could indicate that E2F3 is simply less active when produced in bacteria. To date no one has compared the binding affinity of the different E2F heterodimeric partners to a panel of E2F sites. This could be done using oligonucleotides containing the E2F sites from the cellular promoters listed in Table 3 or by selecting high-affinity binding sites from a collection of random oligonucleotides. Two studies have used cyclic amplification and selection of targets (CAST) techniques to derive an E2F consensus site (CHITTENDEN et al. 1991a; OULETTE et al. 1992). However, both studies relied on the ability of cellular DNA-binding proteins to interact with Rb for the selection of sites from a random population of oligonucleotides. These experiments could also select DNA-binding sites for other cellular proteins that bind to Rb, not just E2F family members; in fact, one study selected an Sp1-binding site (CHITTENDEN et al. 1991a).

The majority of sequences which were isolated in CAST experiments were composed of the consensus TTTTCCCGCCAAAA (OULETTE et al. 1992). Two other classes of sequences were obtained less frequently; these two consensus sequences differed from each other only in the addition of a G (TTTTCCC GCCTTTTTT and TTTTCCCGCGCTTTTTT). The most common sequences cloned in a different study were very similar, although the 5' end differed slightly, perhaps due to the design of the oligonucleotide or to the source of the protein extracts (CHITTENDEN et al. 1991a). It has been suggested that the two different types of E2F sites (containing either A's or T's at the 3' end) may bind different E2F family members (OULETTE et al. 1992). Since binding to Rb was required in the experimental protocol, the E2F family members that bind the selected sites should be E2F1, E2F2, or E2F3, but not E2F4. Although none of the known cellular promoters precisely match the binding sites revealed in the CAST experiments, the promoters that have been shown to be growth regulated via the E2F sites (*dhfr*, *E2F1*, and b-*myb*) all contain 2 to 5 A's 3'of the consensus (creating an overlapping E2F site on the opposite strands of DNA). Perhaps this type of "inverted, overlapping" E2F site specifies binding by a growth-regulatory E2F family member (such as E2F1 or E2F2).

Many studies of protein binding to E2F elements from cellular promoters have been performed using crude protein extracts from cell lines. These studies cannot distinguish among family members unless specific antibodies are used to supershift the complexes. The use of antibodies against specific family members in gel shift assays suggests that the major protein that binds to the *E2* promoter E2F site is not E2F1, E2F2, or E2F3 (WU et al. 1995). Unfortunately, supershifts with family member-specific antibodies have not been performed using E2F sites from different cellular promoters. However, a recent study did compare the sizes of different E2F/protein complexes that can form on the E2F sites from the *dhfr* versus the *E2* promoter (SCHULZE et al. 1994). Two major complexes formed on the *dhfr* promoter-E2F sites, whereas these two, plus an additional complex formed on the *E2* promoter-E2F sites (SCHULZE et al. 1994). Inclusion of antibodies indicated that one of the two common complexes contained Rb and the other p107; the middle complex formed on the *E2* promoter was not recognized by either antibody.

Most studies using gel shift assays have found that the overall amount of binding to E2F sites from various cellular promoters increases about three- to five fold as quiescent cells approach S phase (CHITTENDEN et al. 1993; LI et al. 1994; MUDRYJ et al. 1991). In contrast, PORCU et al. (1994) have shown that binding to the E2F site in the *IGF-1* promoter is high in extracts prepared from quiescent cells and low in extracts prepared from growing cells. It is possible that this difference represents an interaction of a specific E2F family member with this particular E2F site. If so, these results suggest that E2F family members found in quiescent cells (e.g. E2F4), but not E2F family members found in S phase (e.g., E2F1), preferentially recognize the E2F site in the *IGF-1* promoter. Unfortunately, a control reaction using an E2F site that previously has been shown to bind to E2F proteins in S phase extracts was not included in these experiments.

5.2 Can Specificity Be Achieved in Reporter Gene Assays?

The combination of binding sites for different transactivators may determine which E2F family members can most efficiently activate a given promoter. Different E2Fs may contain yet unmapped protein domains located within the nonconserved regions that impart target promoter specificity by virtue of protein-protein interactions. Many E2F-activated promoters also contain Sp1 binding sites (see Table 3), and there are at least four different Sp1 family members (HAGEN et al. 1994; MAJELLO et al. 1994). Perhaps precise combinations of distinct E2F and Sp1 family members interact on certain promoters to impart specificity of activation.

If a particular family member does regulate a specific promoter, one might imagine that differences would be observed if the individual E2F family members were transfected with a promoter-reporter plasmid. The ability of E2F1, E2F2, and E2F3, in combination with DP1 and DP2, to activate a synthetic promoter construct has been examined (WU et al. 1995). E2F1 and E2F2 activate transcription of the reporter promoter to about equal levels, but activation is considerably less with E2F3. In general, all E2Fs are less active when transfected with DP2; these differences are consistent with the reduced DNA-binding activity of the E2F/DP2 heterodimers. Until a larger number of promoters are compared, it is difficult to know whether the differences observed to date simply reflect differences in expression of the E2F proteins in the transfection assays, or whether certain E2F sites are preferentially activated by a distinct subset of the E2F/DP heterodimers.

A caveat to these transfection experiments is that expression of exogenous proteins creates an artificial environment. With a higher concentration of an individual E2F in the cell, an E2F site that is not normally occupied by that factor may be bound. Also, overexpression of one E2F family member might cause the levels of another E2F family member to be increased. The E2F1 promoter contains E2F-binding sites, and transfected E2F1, E2F2, and E2F3 can all stimulate E2F1 promoter activity (HSIAO et al. 1994; JOHNSON et al. 1994b; NEUMAN et al. 1994). Therefore, transfection of different E2F family members might activate transcription of the endogenous E2F1 gene (HSIAO et al. 1994; JOHNSON et al. 1994b; NEUMAN et al. 1994), making it difficult to determine whether, for example, the transfected E2F2 or the endogenous E2F1 activates a reporter construct. Finally, the cellular promoters that are analyzed in reporter assays are not in their normal chromosomal location. If sequences near the promoter, but not contained in the fragment chosen, are critical for determining specificity, false results may be obtained using transfection experiments.

5.3 Regulation of Endogenous Cellular Promoters

The effects of modulating the levels of E2F activity on the regulation of endogenous cellular genes has been investigated in several ways. First, overall E2F activity has been increased by viral oncogenes, either by viral infection of cells or stable transformation of cells with viral oncoproteins. For example, *dhfr* mRNA levels increase after infection with polyoma virus, cytomegalovirus, or adenovirus (see SCHILLING and FARNHAM 1994 for a review). The presumed mechanism for this increase in *dhfr* mRNA is an increase in E2F activity due to sequestration of Rb and p107 by the viral oncoproteins. A more recent study (MUDRAK et al. 1994) analyzed the expression of endogenous cellular genes after induction of expression of a stably integrated T antigen. Induction of T antigen resulted in an increase in E2F activity, demonstrated using a gel shift assay, and an eightfold increase in the levels of thymidine kinase mRNA; however, very little increase in the levels of *dhfr* or DNA polymerase-α mRNAs was observed. *E2F1* mRNA levels decreased, indicating that the increased E2F DNA-binding activity is likely due to changes in levels of a different E2F family member. One interpretation of this experiment is that T antigen increases the amount or activity of an E2F family member that can regulate the thymidine kinase promoter, but not the *dhfr*, DNA polymerase-α, or *E2F1* promoters. However, it is not known what family member is responsible for the increased E2F DNA-binding activity seen after expression of T antigen.

Several groups have examined changes in levels of endogenous cellular mRNAs after overexpression of *E2F1*. In one study colonies were obtained containing stably integrated *E2F1* that was expressed from a strong viral promoter (SINGH et al. 1994). Although levels of exogenous *E2F1* were very high (at least ten times the amount of *E2F1* that is normally present), the levels of the endogenous *E2F1* mRNA were not increased. The mRNA levels of other potential target genes, c-*myc* and *dhfr*, were increased about twofold, whereas a sixfold increase was seen in thymidine kinase mRNA. Although these mRNAs were analyzed by Northern blots, and thus quantitation is not very accurate, these data suggest that increasing the level of E2F1 is not sufficient to increase transcription from the *E2F1*, *dhfr*, or c-*myc* promoters. Because the levels of other E2F family members were not examined, it cannot be determined whether the increased levels of thymidine kinase mRNA are a direct consequence of the expression of exogenous E2F1 or an indirect effect of the change in levels of a different family member. In support of the idea that levels of a different E2F family member can be increased by overexpression of E2F1, cells from morphologically transformed foci induced by overexpression of E2F1 show increased E2F-DNA-binding activity mediated by a protein that is not recognized by an antibody specific to human E2F1 (JOHNSON et al. 1994a). However, it cannot be ruled out that the E2F-binding activity was mediated by E2F1; it is possible that antibody binding could have been blocked by other proteins present in the complex. A second investigation into the regulation of endogenous cellular genes by E2F1 was performed by infecting quiescent cells with an adenoviral vector expressing E2F1 (J.R. Nevins, personal communication). This study found that the levels of *E2F1*, cyclin A, b-*myb*, and cyclin E mRNAs were increased, but the levels of *dhfr*, DNA polymerase-α, and thymidine kinase mRNAs did not change. It is unclear why overexpression of E2F1 can increase levels of different endogenous mRNAs in different investigations. However, it is possible that differences in cell types or levels of expression of E2F1 alter the outcome of the experiments. A caveat with all of these experiments concerns the deregulated expression of the E2F family member. Although examination of endogenous genes is one step closer to true regulation than is the use of a promoter reporter assay, it is quite possible that correct specificity is observed only if the E2F family member is expressed at normal levels and at the correct time in the cell cycle. Experiments for further exploration of this topic are presented by FARNHAM (this volume).

6 Summary

E2F is a heterodimer composed of two partners, such as E2F1 and DP1. Although E2F1 can bind DNA as a homodimer and increase promoter activity, optimal DNA-binding and transcriptional activity occurs in the heterodimeric form. A model (Fig. 3) for the involvement of E2F activity in cell growth control that incorporates viral oncoproteins, positive regulators of cell growth (cyclins) and negative regulators of cell growth (tumor suppressor proteins) can now be advanced. Each aspect of this model is addressed in subsequent chapters of this book. It is likely that binding of growth-suppressing proteins, such as Rb, can inhibit the transactivation potential of E2F1, either by blocking the interaction of E2F1 with a separate component of the transcription complex or by bringing a repressor domain to the transcription complex (FLEMINGTON et al. 1993; HELIN et al. 1993; WEINTRAUB et al. 1992; ZAMANIAN and LA THANGUE 1993; ZHU et al. 1993). Phosphorylation or sequestration of Rb by viral oncoproteins can free E2F. The influence of viral oncoproteins on E2F activity and the regulation of the different E2F complexes is the focus of the contributions by COBRINIK and by CRESS and NEVINS. The interaction of the free E2F induces a bend in the DNA that may also play a role in transactivation, perhaps by bringing proteins (such as an Sp1 or CCAAT family member) separated by distance on the promoter DNA into contact (HUBER et al. 1994). Because E2F

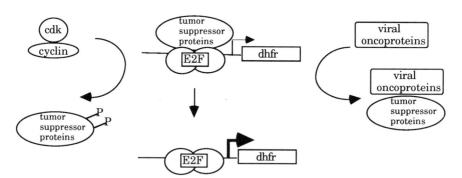

Fig. 3. Model for regulation of E2F-mediated transcription. E2F-mediated transcriptional activation of cellular genes such as *dhfr* can be increased by the action of growth-promoting proteins such as cyclins (see COBRINIK, this volume) and viral oncoproteins (see CRESS and NEVINS, this volume)

target genes encode proteins critical for cell growth, deregulation of E2F activity can have severe consequences, such as apoptosis or uncontrolled proliferation. The effect of altered expression of E2F activity on the cell cycle and on tumorigenicity is the focus of the contribution by ADAMS and KAELIN. Finally, a comparison of E2F to the genetically well-characterized factors that regulate G_1/S phase transcription in yeast is the subject of the chapter by BREEDEN. This volume concludes with FARNHAM's summary of the rapid gains in knowledge concerning the E2F gene family that have been made in the past several years and provides a series of questions and lines of investigation that will be the focus of future studies.

Acknowledgments. Work in this laboratory was supported by U.S. Public Health Service grants CA45240 and CA59524 from the National Institutes of Health to P.J.F. J.E.S was supported in part by Public Health Service training grant CA09135 from the National Institutes of Health and in part as a Cremer Scholar. We thank Joe Nevins, Renee Bernards, Gary Lyons, Chin-Lee Wu, Ed Harlow, Judith Campisi, Michael Greenberg, Amy Yee, Jackie Lees, Yash Vaishnav, Kuang-Ming Hsiao, Christopher Fry, Stephanie McMahon and Kathrine Wright for communicating unpublished results, and members of the Farnham laboratory, especially Erika Meyer, for critical reading of this manuscript.

References

Bagchi S, Raychaudhuri P, Nevins JR (1990) Adenovirus E1A proteins can dissociate heteromeric complexes involving the E2F transcription factor: a novel mechanism for trans-activation. Cell 62: 659–669

Bandara LR, Adamczewski JP, Hunt T, La Thangue NB (1991) Cyclin A and the retinoblastoma gene product complex with a common transcription factor. Nature 352: 249–251

Bandara LR, Buck VM, Zamanian M, Johnston LH, La Thangue NB (1993) Functional synergy between DP-1 and E2F-1 in the cell cycle-regulating transcription factor DRTF1/E2F. EMBO J 12: 4317–4324

Bandara LR, Lam EW, Sørenson TS, Zamanian M, Girling R, La Thangue NB (1994) DP-1: a cell cycle-regulated and phosphorylated component of transcription factor DRTF1/E2F which is functionally important for recognition by pRb and the adenovirus E4 orf 6/7 protein. EMBO J 13: 3104–3114

Beijersbergen RL, Kerkhoven RM, Zhu L, Carlée L, Voorhoeve PM, Bernards R (1994) E2F4, a new member of the E2F gene family, has oncogenic activity and associates with p107 in vivo. Genes Dev 8: 2680–2690

Bennett LM, Farnham PJ, Drinkwater NR (1995) Strain-specific differences in DNA synthesis and gene expression in the regenerating livers of C57BL/6J and C3H/HeJ mice. Mol Carcinogenesis (in press)

Blake MC, Azizkhan JC (1989) Transcription factor E2F is required for efficient expression of the hamster dihydrofolate reductase gene in vitro and in vivo. Mol Cell Biol 9: 4994–5002

Chang Z, Cheng S (1993) Constitutive overexpression of DNA binding activity to the distal CCAAT box of human thymidine kinase promoter in human tumor cell lines. Cancer Res 53: 3253–3256

Chellappan SP, Hiebert SW, Mudryj M, Horowitz JM, Nevins JR (1991) The E2F transcription factor is a cellular target for the RB protein. Cell 65: 1053–1061

Chellappan SP, Kraus VB, Kroger B, Munger K, Howley PM, Phelps WC, Nevins JR (1992) Adenovirus E1A, simian virus 40 tumor antigen, and human papillomaviurs E7 protein share the capacity to disrupt the interaction between transcription factor E2F and the retinoblastoma gene product. Proc Natl Acad Sci USA 89: 4549–4553

Chittenden T, Livingston DM, Kaelin WG Jr (1991a) The T/E1A-binding domain of the retinoblastoma product can interact selectively with a sequence-specific DNA-binding protein. Cell 65: 1073–1082

Chittenden T, Livingston DM, Kaelin WG (1991b) RB associates with an E2F-like sequence-specific DNA binding protein. Cold Spring Harb Symp Quant Biol 56: 187–195

Chittenden T, Livingston DM, DeCaprio JA (1993) Cell cycle analysis of E2F in primary human T cells reveals novel E2F complexes and biochemically distinct forms of free E2F. Mol Cell Biol 13: 3975–3983

Cress WD, Johnson DG, Nevins JR (1993) A genetic analysis of the E2F1 gene distinguishes regulation by Rb, p107, and adenovirus E4. Mol Cell Biol 13: 6314–6325

Dalton S (1992) Cell cycle regulation of the human cdc2 gene. EMBO J 11: 1797–1804

DePinho RA, Legouy E, Feldman LB, Kohl NE, Yancopoulos GD, Alt FW (1986) Structure and expression of the murine N-*myc* gene. Proc Natl Acad Sci USA 83: 1827–1831

Dou Q-P, Zhao S, Levin AH, Wang J, Helin K, Pardee AB (1994) G₁/S-regulated E2F-containing protein complexes bind to the mouse thymidine kinase gene promoter. J Biol Chem 269: 1306–1313

Dutta A, Stoeckle MY, Hanafusa H (1990) Serum and v-src increase the level of a CCAAT-binding factor required for transcription from a retroviral long terminal repeat. Genes Dev 4: 243–254

Dynlacht BD, Brook A, Dembski M, Yenush L, Dyson N (1994a) DNA-binding and trans-activation properties of *Drosophila* E2F and DP proteins. Proc Natl Acad Sci USA 91: 6359–6363

Dynlacht BD, Flores O, Lees JA, Harlow E (1994b) Differential regulation of E2F trans-activation by cyclin/cdk2 complexes. Genes Dev 8: 1772–1786

Dyson N, Dembski M, Fattaey A, Ngwu C, Ewen M, Helin K (1993) Analysis of p107-associated proteins: p107 associates with a form of E2F that differs from pRB-associated E2F-1. J Virol 67: 7641–7647

Ewen ME, Faha B, Harlow E, Livingston DM (1992) Interaction of p107 with cyclin A independent of complex formation with viral oncoproteins. Science 255: 85–90

Fagan R, Flint KJ, Jones N (1994) Phosphorylation of E2F-1 modulates its interaction with the retinoblastoma gene product and the adenoviral E4 19 kDa protein. Cell 78: 799–811

Farnham PJ, Schimke RT (1986) Murine dihydrofolate reductase transcripts through the cell cycle. Mol Cell Biol 6: 365–371

Farnham PJ, Slansky JE, Kollmar R (1993) The role of E2F in the mammalian cell cycle. Biochim Biophys Acta 1155: 125–131

Fisher DE, Parent LA, Sharp PA (1993) High affinity DNA-binding myc analogs: recognition by an α helix. Cell 72: 467–476

Flemington EK, Speck SH, Kaelin WG Jr (1993) E2F1-mediated trans-activation is inhibited by complex formation with the retinoblastoma susceptibility gene product. Proc Natl Acad Sci USA 90: 6914–6918

Furukawa Y, Terui Y, Sakoe K, Ohta M, Saito M (1994) The role of cellular transcription factor E2F in the regulation of cdc2 mRNA expression and cell cycle control of human hematopoietic cells. J Biol Chem 269: 26249–26258

Gill RM, Hamel PA, Jiang Z, Zacksenhaus E, Gallie BL, Phillips RA (1994) Characterization of the human RB1 promoter and of elements involved in transcriptional regulation. Cell Growth Differ 5: 467–474

Ginsberg D, Vairo G, Chittenden T, Xiao Z-X, Xu G, Wydner KL, DeCaprio JA, Lawrence JB, Livingston DM (1994) E2F-4, a new member of the E2F transcription factor family, interacts with p107. Genes Dev 8: 2665–2679

Girling R, Partridge JF, Bandara LR, Burden N, Totty NF, Hsuan JJ, La Thangue NB (1993) A new component of the transcription factor DRTF1/E2F. Nature 362: 83–87 (authors' correction: 365: 468)

Goldsmith ME, Beckman CA, Cowan KH (1986) 5' Nucleotide sequences influence serum-modulated expression of a human dihydrofolate reductase minigene. Mol Cell Biol 6: 878–886

Goodrich DW, Wang NP, Qian Y-W, Lee EY-H, Lee W-H (1991) The retinoblastoma gene product regulates progression through the G₁ phase of the cell cycle. Cell 67: 293–302

Hagemeier C, Cook A, Kouzarides T (1993) The retinoblastoma protein binds E2F residues required for activation in vivo and TBP binding in vitro. Nucleic Acids Res 21: 4998–5004

Hagen G, Muller S, Beato M, Suske G (1994) Sp1-mediated transcriptional activation is repressed by Sp3. EMBO J 13: 3843–3851

Hara E, Okamoto S, Nakada S, Taya Y, Sekiya S, Oda K (1993) Protein phosphorylation required for the formation of E2F complexes regulates N-myc transcription during differentiation of human embryonal carcinoma cells. Oncogene 8: 1023–1032

Hardy S, Shenk T (1989) E2F from adenovirus-infected cells binds cooperatively to DNA containing two properly oriented and spaced recognition sites. Mol Cell Biol 9: 4495–4506

Helin K, Lees JA, Vidal M, Dyson N, Harlow E, Fattaey A (1992) A cDNA encoding a pRB-binding protein with properties of the transcription factor E2F. Cell 70: 337–350

Helin K, Harlow E, Fattaey A (1993a) Inhibition of E2F-1 transactivation by direct binding of the retinoblastoma protein. Mol Cell Biol 13: 6501–6508

Helin K, Wu C-L, Fattaey AR, Lees JA, Dynlacht BD, Ngwu C, Harlow E (1993b) Heterodimerization of the transcription factors E2F-1 and DP1 leads to cooperative trans-activation. Genes Dev 7: 1850–1861

Hengstschläger M, Mudrak I, Wintersberger E, Wawra E (1994) A common regulation of genes encoding enzymes of the deoxynucleotide metabolism is lost after neoplastic transformation. Cell Growth Differ 5: 1389–1394

Hiebert SW, Lipp M, Nevins JR (1989) E1A-dependent trans-activation of the human MYC promoter is mediated by the E2F factor. Proc Natl Acad Sci USA 86: 3594–3598

Hiebert SW, Blake M, Azizkhan J, Nevins JR (1991) Role of E2F transcription factor in E1A-mediated trans activation of cellular genes. J Virol 65: 3547–3552

Hiebert SW, Chellappan SP, Horowitz JM, Nevins JR (1992) The interaction of RB with E2F coincides with an inhibition of the transcriptional activity of E2F. Genes Dev 6: 177–185

Hsiao K-M, McMahon SL, Farnham PJ (1994) Multiple DNA elements are required for the growth regulation of the mouse E2F1 promoter. Genes Dev 8: 1526–1537

Huber HE, Edwards G, Goodhart PJ, Patrick DR, Huang PS, Ivey-Hoyle M, Barnett SF, Oliff A, Heimbrook DC (1993) Transcription factor E2F binds as a heterodimer. Proc Natl Acad Sci USA 90: 3525–3529

Huber HE, Goodhart PJ, Huang PS (1994) Retinoblastoma protein reverses DNA bending by transcription factor E2F. J Biol Chem 269: 6999–7005

Ivey-Hoyle M, Conroy R, Huber HE, Goodhart PJ, Oliff A, Heimbrook DC (1993) Cloning and characterization of E2F-2, a novel protein with the biochemical properties of transcription factor E2F. Mol Cell Biol 13: 7802–7812

Johnson DG, Cress WD, Jakoi L, Nevins JR (1994a) Oncogenic capacity of the E2F1 gene. Proc Natl Acad Sci USA 91: 12823–12827

Johnson DG, Ohtani K, Nevins JR (1994b) Autoregulatory control of E2F1 expression in response to positive and negative regulators of cell cycle progression. Genes Dev 8: 1514–1525

Jordan KL, Haas AR, Logan TJ, Hall DJ (1994) Detailed analysis of the basic domain of the E2F1 transcription factor indicates that it is unique among bHLH proteins. Oncogene 9: 1177–1185

Kaelin WG Jr, Krek W, Sellers WR, DeCaprio JA, Ajchenbaum F, Fuchs CS, Chittenden T, Li Y, Farnham PJ, Blanar MA, Livingston DM, Flemington EK (1992) Expression cloning of a cDNA encoding a retinoblastoma-binding protein with E2F-like properties. Cell 70: 351–364

Kim S-J, Onwuta US, Lee YI, Li R, Botchan MR, Robbins PD (1992) The retinoblastoma gene product regulates Sp1-mediated transcription. Mol Cell Biol 12: 2455–2463

Kim YK, Lee AS (1991) Identification of a 70-base-pair cell cycle regulatory unit within the promoter of the human thymidine kinase gene and its interaction with cellular factors. Mol Cell Biol 10: 2296–2302

Kim YK, Lee AS (1992) Identification of a protein-binding site in the promoter of the human thymidine kinase gene required for the G1-S-regulated transcription. J Biol Chem 267: 2723–2727

Kovesdi I, Reichel R, Nevins JR (1986) Identification of a cellular transcription factor involved in E1A transactivation. Cell 45: 219–228

Krek W, Livingston DM, Shirodkar S (1993) Binding to DNA and the retinoblastoma gene product promoted by complex formation of different E2F family members. Science 262: 1557–1560

Krek W, Ewen ME, Shirodkar S, Arany Z, Kaelin WG Jr, Livingston DM (1994) Negative regulation of the growth-promoting transcription factor E2F-1 by a stably bound cyclin A-dependent protein kinase. Cell 78: 161–172

La Thangue N, Rigby PW (1987) An adenovirus E1A-like transcription factor is regulated during the differentiation of murine embryonal carcinoma stem cells. Cell 49: 507–513

Lam EW-F, Watson RJ (1993) An E2F-binding site mediates cell-cycle regulated repression of mouse B-myb transcription. EMBO J 12: 2705–2713

Lam EW-F, Morris JDH, Davies R, Crook T, Watson RJ, Vousden KH (1994) HPV16 E7 oncoprotein deregulates B-myb expression: correlation with targeting of p107/E2F complexes. EMBO J 13: 871–878

Lee Y, Yano M, Liu S, Matsunaga E, Johnson PF, Gonzalez FJ (1994) A novel cis-acting element controlling the rat CYP2D5 gene and requiring cooperativity between C/EBPβ and an Sp1 factor. Mol Cell Biol 14: 1383–1394

Lees E, Faha B, Dulic V, Reed SI, Harlow E (1992) Cyclin E/cdk2 and cyclin A/cdk2 kinases associate with p107 and E2F in a temporally distinct manner. Genes Dev 6: 1874–1885

Lees JA, Saito M, Vidal M, Valentine M, Look T, Harlow E, Dyson N, Helin K (1993) The retinoblastoma protein binds to a family of E2F transcription factors. Mol Cell Biol 13: 7813–7825

Li L, Naeve GS, Lee AS (1993) Temporal regulation of cyclin A-p107 and p33cdk2 complexes binding to a human thymidine kinase promoter element important for G_1-S phase transcriptional regulation. Proc Natl Acad Sci USA 90: 3554–3558

Li Y, Slansky JE, Myers DJ, Drinkwater NR, Kaelin WG, Farnham PJ (1994) Cloning, chromosomal location, and characterization of mouse E2F1. Mol Cell Biol 14: 1861–1869

Mai B, Lipp M (1993) Identification of a protein from Saccharomyces cerevisiae with E2F-like DNA-binding and transactivating properties. FEBS Lett 321: 153–158

Majello B, De Luca P, Hagen G, Suske G, Lania L (1994) Different members of the Sp1 multigene family exert opposite transcriptional regulation of the long terminal repeat of HIV-1. Nucleic Acids Res 22: 4914–4921

Malhotra P, Manohar CF, Swaminathan S, Toyama R, Dhar R, Reichel R, Thimmapaya B (1993) E2F site activates transcription in fission yeast Schizosaccharomyces pombe and binds to a 30-KDa transcription factor. J Biol Chem 268: 20392–20401

Mariani BD, Slate DL, Schimke RT (1981) S phase-specific synthesis of dihydrofolate reductase in Chinese hamster ovary cells. Proc Natl Acad Sci USA 78: 4985–4989

Martinelli R, Heintz N (1994) HITF2A, the large subunit of a heterodimeric, glutamine-rich CCAAT-binding transcription factor involved in histone H1 cell cycle regulation. Mol Cell Biol 14: 8322–8332

Means AL, Slansky JE, McMahon SL, Knuth MW, Farnham PJ (1992) The HIP1 binding site is required for growth regulation of the dihydrofolate reductase gene promotor. Mol Cell Biol 12: 1054–1063

Melamed D, Tiefenbrun N, Yarden A, Kimchi A (1993) Interferons and interleukin-6 suppress the DNA-binding activity of E2F in growth-sensitive hematopoietic cells. Mol Cell Biol 13: 5255–5265

Miltenberger RJ, Farnham PJ, Smith DE, Stommel JM, Cornwell MM (1995) v-Raf activates trans-cription of growth-responsive promoters via GC-rich sequences that bind the transcription factor Sp1. Cell Growth Differ 6: 549–559

Mitchell PJ, Carothers AM, Han JH, Harding JD, Kas E, Venolia L, Chasin LA (1986) Multiple transcription start sites, DNase I-hypersensitive sites, and an opposite-strand exon in the 5' region of the CHO dhfr gene. Mol Cell Biol 6: 425–440

Moberg KH, Logan TJ, Tyndall WA, Hall DJ (1992) Three distinct elements within the murine c-myc promoter are required for transcription. Oncogene 7: 411–421

Mudrak I, Ogris E, Rotheneder H, Wintersberger E (1994) Coordinated trans activation of DNA synthesis- and precursor-producing enzymes by polyomavirus large T antigen through interaction with the retinoblastoma protein. Mol Cell Biol 14: 1886–1892

Mudryj M, Hiebert SW, Nevins JR (1990) A role for the adenovirus inducible E2F transcription factor in a proliferation dependent signal transduction pathway. EMBO J 9: 2179–2184

Mudryj M, Devoto SH, Hiebert SW, Hunter T, Pines J, Nevins JR (1991) Cell cycle regulation of the E2F transcription factor involves an interaction with cyclin A. Cell 65: 1243–1253

Neuman E, Flemington EK, Sellers WR, Kaelin WG Jr (1994) Transcription of the E2F-1 gene is rendered cell cycle dependent by E2F DNA-binding sites within its promoter. Mol Cell Biol 14: 6607–6615 (authors' correction: 15: 4660)

Nevins JR (1992) Transcriptional activation by viral regulatory proteins. Trends Biochem Sci 16: 435–439

Ogris E, Rotheneder H, Mudrak I, Pichler A, Wintersberger E (1993) A binding site for transcription factor E2F is a target for trans activation of murine thymidine kinase by polyomavirus large T antigen and plays an important role in growth regulation of the gene. J Virol 67: 1765–1771

Ohtani K, Nevins JR (1994) Functional properties of a Drosophila homolog of the E2F1 gene. Mol Cell Biol 14: 1603–1612

Oswald F, Lovec H, Möröy T, Lipp M (1994) E2F-dependent regulation of human MYC: trans-activation by cyclins D1 and A overrides tumour suppressor protein functions. Oncogene 9: 2029–2036

Oulette MM, Chen J, Wright WE, Shay JW (1992) Complexes containing the retinoblastoma gene product recognize different DNA motifs related to the E2F binding site. Oncogene 7: 1075–1081

Pang JH, Chen KY (1993) A specific CCAAT-binding protein, CBP/tk, may be involved in the regulation of thymidine kinase gene expression in human IMR-90 diploid fibroblasts during senescence. J Biol Chem 268: 2909–2916

Park K, Choe J, Osifchin NE, Templeton DJ, Robbins PD, Kim S-J (1994) The human retinoblastoma susceptibility gene promoter is positively autoregulated by its own product. J Biol Chem 269: 6083–6088

Pearson BE, Nasheuer H-P, Wang TS-F (1991) Human DNA polymerase α gene: sequences controlling expression in cycling and serum-stimulated cells. Mol Cell Biol 11: 2081–2095

Peeper DS, Keblusek P, Helin K, Toebes M, van der Eb AJ, Zantema A (1995) Phosphorylation of a specific cdk site in E2F-1 affects its electrophoretic mobility and promotes pRB-binding in vitro. Oncogene 10: 39–48

Philpott A, Friend SH (1994) E2F and its developmental regulation in Xenopus laevis. Mol Cell Biol 14: 5000–5009

Porcu P, Ferber A, Pietrzkowski Z, Roberts CT, Adamo M, LeRoith D, Baserga R (1992) The growth-stimulatory effect of simian virus 40 T antigen requires the interaction of insulinlike growth factor 1 with its receptor. Mol Cell Biol 12: 5069–5077

Porcu P, Grana X, Li S, Swantek J, De Luca A, Giordano A, Baserga R (1994) An E2F binding sequence negatively regulates the response of the insulin-like growth factor 1 (IGF-I) promoter to simian virus 40 T antigen and to serum. Oncogene 9: 2125–2134

Qin XQ, Chittenden T, Livingston DM, Kaelin WG (1992) Identification of a growth suppression domain within the retinoblastoma gene product. Genes Dev 6: 953–964

Roussel MF, Davis JN, Cleveland JL, Ghysdael J, Hiebert SW (1994) Dual control of myc expression through a single DNA binding site targeted by ets family proteins and E2F-1. Oncogene 9: 405–415

Saito M, Helin K, Valentine MB, Griffith BB, Willman CL, Harlow E, Look AT (1995) Amplification of the E2F1 transcription factor gene in the HEL erythroleukemia cell line. Genomics 25: 130–138

Santiago C, Collins M, Johnson LF (1984) In vitro and in vivo analysis of the control of dihydrofolate reductase gene transcription in serum-stimulated mouse fibroblasts. J Cell Physiol 118: 79–86

Sardet C, Vidal M, Cobrinik D, Geng Y, Onufryk C, Chen A, Weinberg RA (1995) E2F-4 and E2F-5, two members of the E2F family, are expressed in the early phases of the cell cycle. Proc Natl Acad Sci USA 92: 2403–2407

Satterwhite DJ, Aakre ME, Gorska AE, Moses HL (1994) Inhibition of cell growth by TGFβ1 is associated with inhibition of B-myb and cyclin A in both BALB/MK and MV1Lu cells. Cell Growth Differ 5: 789–799

Schilling LJ, Farnham PJ (1994) Transcriptional regulation of the dihydrofolate reductase/rep-3 locus. Crit Rev Euk Gene Exp 4: 19–53

Schilling LJ, Farnham PJ (1995) The bidirectionally transcribed dhfr and rep-3a promoters are growth regulated by distinct mechanisms. Cell Growth Differ 6: 541–548

Schulze A, Zerfaß K, Spitkovsky D, Henglein B, Jansen-Dürr P (1994) Activation of the E2F transcription factor by cyclin D1 is blocked by p16INK4, the product of the putative tumor suppressor gene MTS1. Oncogene 9: 3475–3482

Schwarz JK, Devoto SH, Smith EJ, Chellappan SP, Jakoi L, Nevins JR (1993) Interactions of the p107 and Rb proteins with E2F during the cell proliferation response. EMBO J 12: 1013-1021

Shan B, Zhu X, Chen P-L, Durfee T, Yang Y, Sharp D, Lee W-H (1992) Molecular cloning of cellular genes encoding retinoblastoma-associated proteins: identification of a gene with properties of the transcription factor E2F. Mol Cell Biol 12: 5620–5631

Shan B, Chang CY, Jones D, Lee WH (1994) The transcription factor E2F-1 mediates the autoregulation of RB gene expression. Mol Cell Biol 14: 299–309

Shimada T, Inokuchi K, Nienhuis AW (1986) Chromatin structure of the human dihydrofolate reductase gene promoter. J Biol Chem 261: 1445–1452

Singh P, Wong SH, Hong W (1994) Overexpression of E2F-1 in rat embryo fibroblasts leads to neoplastic transformation. EMBO J 13: 3329–3338

Slansky JE, Farnham PJ (1993) The role of the transcription factor E2F in the growth regulation of DHFR. In: Hu VW (ed) The cell cycle: regulators, targets, and clinical applications. Plenum, New York

Slansky JE, Li Y, Kaelin WG, Farnham PJ (1993) A protein synthesis-dependent increase in E2F1 mRNA correlates with growth regulation of the dihydrofolate reductase promoter. Mol Cell Biol 13: 1610–1618 (author's correction: 13: 7201)

Song JJ, Walker S, Chen E, Johnson EEI, Spychala J, Gribbin T, Mitchell BS (1993) Genomic structure and chromosomal location of the human deoxycytidine kinase gene. Proc Natl Acad Sci USA 90: 431–434

Takehana K, Nakada S, Hara E, Taya Y, Sekiya S, Oda K (1991) Interaction of nuclear factors with the regulatory region of the N-myc gene during differentiation of human embryonal carcinoma cells. Gene 103: 219–225

Tevosian SS, Paulson KE, Bronson R, Yee AS (1995) Expression of the E2F-1/DP-1 transcription factor in murine development (submitted)

Thalmeier K, Synovzik H, Mertz R, Winnacker E-L, Lipp M (1989) Nuclear factor E2F mediates basic transcription and trans-activation by E1a of the human MYC promoter. Genes Dev 3: 527–536

Udvadia AJ, Rogers KT, Higgins PDR, Murata Y, Martin KH, Humphrey PA, Horowitz JM (1993) Sp-1 binds promoter elements regulated by the RB protein and Sp-1-mediated transcription is stimulated by RB coexpression. Proc Natl Acad Sci USA 90: 3265–3269

Wade M, Kowalik TF, Mudryj M, Huang E-S, Azizkhan JC (1992) E2F mediates dihydrofolate reductase promoter activation and multiprotein complex formation in human cytomegalovirus infection. Mol Cell Biol 12: 4364–4374

Watson RJ, Dyson PJ, McMahon J (1987) Multiple c-myb transcript cap sites are variously utilized in cell of mouse haemopoietic origin. EMBO J 6: 1643–1651

Weintraub SJ, Prater CA, Dean DC (1992) Retinoblastoma protein switches the E2F site from a positive to negative element. Nature 358: 259–261

Whyte P, Buchkovich KJ, Horowitz JM, Friend SH, Raybuck M, Weinberg RA, Harlow E (1988) Association between an oncogene and an antioncogene: the adenovirus E1A proteins bind to the retinoblastoma gene product. Nature 334: 124–129

Wu CL, Zukerberg LR, Ngwu S, Harlow E, Lees J (1995) In vivo association of E2F and DP family proteins. Mol Cell Biol 15: 2536–2546

Xu M, Sheppard KA, Peng CY, Yee AS, Piwnica-Worms H (1994) Cyclin A/CDK2 binds directly to E2F-1 and inhibits the DNA-binding activity of E2F-1/DP-1 by phosphorylation. Mol Cell Biol 14: 8420–8431

Zamanian M, La Thangue NB (1993) Transcriptional repression by the Rb-related protein p107. Mol Biol Cell 4: 389–396

Zhou W, Takuwa N, Kumada M, Takuwa Y (1994) E2F1, b-myb and selective members of cyclin/cdk subunits are targets for protein kinase C-mediated bimodal growth regulation in vascular endothelial cells. Biochem Biophys Res Commun 199: 191–198

Zhu L, van den Heuvel S, Helin K, Fattaey A, Ewen M, Livingston D, Dyson N, Harlow E (1993) Inhibition of cell proliferation by p107, a relative of the retinoblastoma protein. Genes Dev 7: 1111–1125

Regulatory Interactions Among E2Fs
and Cell Cycle Control Proteins

D. COBRINIK

Whitehead Institute for Biomedical Research, 9 Cambridge Center, Cambridge, MA 02142, USA
Present Address: Division of Medical Oncology, Department of Medicine, Columbia University, College of Physicians and Surgeons, 630 West 168th Street, New York, NY 10032, USA

1 Introduction

The last few years have seen an explosive growth in research on E2F transcription factors. Originally characterized as cellular proteins that are critical for adenovirus gene expression and replication, E2Fs have come to be seen as central players in the control of the animal cell cycle. This view comes about largely as the result of the remarkable finding that E2Fs are stably bound and regulated by the pRB tumor suppressor protein. The direct connections between E2Fs and the cell cycle control machinery go far beyond the pRB:E2F association, however, and are now realized to comprise a network of interactions among members of the pRB, E2F, cyclin, and cyclin-dependent kinase (cdk) protein families.

Understanding of E2F multiprotein complexes derives from a convergence of research in the cell cycle, tumor suppressor, and DNA tumor virus fields. This convergence stems from evidence that the binding of DNA tumor virus onco-proteins to pRB overcomes pRB's tumor-suppressing ability and contributes to cell transformation. The inability of DNA tumor virus oncoproteins to bind to inactive mutant forms of pRB that are found in retinoblastoma or other tumor cells suggested that these oncoproteins bind to a pRB "pocket" region that is critical to pRB's growth inhibiting capability. One way in which this binding was proposed to overcome pRB's growth-inhibition is through the displacement of cellular growth-promoting proteins from the pRB pocket domain. We now realize that E2Fs are a group of such proteins whose displacements from the pocket regions of pRB as well as pRB-related proteins deregulates cell growth.

1.1 Grind and Bind Finds: What the Gel Shifts Tell Us

The E2F transcription factor was first identified as an E1A-inducible cellular activity that binds an enhancer element in the adenovirus E2 promoter and stimulates E2 gene expression (KOVESDI et al. 1986, 1987). E1A-inducible E2F was shortly thereafter found to stimulate cellular genes acting through enhancer elements that are related to the E2 enhancer (TTTCGCGC) DNA sequence (BLAKE and AZIZKHAN 1989; HIEBERT et al. 1989; THALMEIER et al. 1989). Although early gel shift analyses with HeLa cell extracts suggested that E2F binds the E2 enhancer as a single discrete species (KOVESDI et al. 1986), multiple complexes with E2F binding specificity were detected in other cell types (LA THANGUE and RIGBY 1987; BAGCHI et al. 1990).

In investigating E2F complexes in a variety of cell types, BAGCHI et al. (1990) demonstrated that the diverse E2F DNA binding activities detected in gel shifts are composed of multiprotein complexes that can be disrupted by detergent to release a "free E2F" species. Interestingly, these authors also found that E2F is released from multiprotein complexes when cells are infected with adenovirus.

BAGCHI et al. (1990) then demonstrated that multiprotein E2F complexes can be disrupted in vitro solely through the addition of adenovirus E1A protein to uninfected cell lysates. Moreover, this disruption depends upon the E1A conserved region (CR)2 domain (BAGCHI et al. 1990; RAYCHAUDHURI et al. 1991) that is critical both for E1A to transform cells and for E1A to bind the pRB, p107, and p130 proteins (YEE and BRANTON 1985; HARLOW et al. 1986; WHYTE et al. 1988, 1989; EGAN et al. 1989; BARBEAU et al. 1994). On this basis it was proposed that E2F multiprotein complexes contain pRB or related proteins, that E1A's binding to the latter proteins releases free E2F, and that the release of free E2F contributes to E1A's transforming capability. As described here and in related articles in this volume, the accuracy of this scenario has been validated.

2 The pRB:E2F Interaction

2.1 Identification of pRB:E2F Complexes

Several groups simultaneously reported an interaction of pRB with E2F transcription factors (BAGCHI et al. 1991; BANDARA et al. 1991; CHELLAPPAN et al. 1991; CHITTENDEN et al. 1991). The most direct indications of this interaction were the demonstrations that (a) pRB antibody disrupts a specific E2F gel shift complex, (b) the same complex that is disrupted by pRB antibody can also be disrupted by E1A in a manner that depends upon E1A's pRB-binding domain, (c) pRB copurifies with E2F activity in oligonucleotide affinity chromatography, and (d) E2F coimmunoprecipitates with pRB (CHELLAPPAN et al. 1991). Importantly, the ability to detect pRB:E2F complexes with both the gel shift and affinity chromatography analyses indicates that pRB binding to E2F does not preclude E2F's binding to DNA. While these observations were first made with extracts of U937 leukemia cells, similar approaches have detected pRB:E2F interactions in a wide variety of other cell types.

Alternative strategies for detecting the pRB–E2F interaction have revealed additional properties of pRB:E2F complexes. CHITTENDEN et al. (1991) detected the pRB–E2F interaction when they used a polymerase chain reaction based approach to identify transcription factor targets of pRB. These authors found that E2F DNA binding sites were by far the major sequences bound by pRB-associating proteins in a WERI-Rb27 retinoblastoma cell extract. This finding suggested that, at least in the WERI-Rb27 cell extract, E2Fs are the most abundant pRB-binding proteins that are also sequence-specific DNA binding proteins.

pRB's interaction with E2F has also been detected through reconstitution studies. For example, bacteria-produced proteins consisting of a portion of pRB fused to glutathione S-transferase were found in gel shift experiments to bind and supershift free E2F species present in extracts as diverse as mouse F9 embryonal carcinoma cells (BANDARA et al. 1991) and *Xenopus* eggs (PHILPOTT and FRIEND 1994).

In other gel shift experiments, however, addition of purified pRB *inhibited* the binding of purified E2F to its DNA recognition sequence (BAGCHI et al. 1991; HIEBERT et al. 1992; Cobrinik, unpublished data). Apparently, one or more components present in cell extracts is necessary for purified pRB:E2F complexes to bind DNA in vitro (HIEBERT et al. 1992; RAY et al. 1992; and Cobrinik, unpublished data). RAY et al. (1992) identified a 60-kDa protein that promotes DNA binding by pRB:E2F complexes; yet this in vitro DNA binding can also be restored by spermine (COBRINIK et al. 1993 and unpublished data), which is a ubiquitous component of eukaryotic cells. Thus, the physiologic relevance of pRB's "E2F-inhibitor" activity is unclear.

2.2 Characteristics of pRB:E2F Complexes

The identification of E2F as a prominent pRB-binding protein suggested that the pRB–E2F interaction contributes to pRB's tumor suppressing capability. This view is supported by the inability of mutant pRB molecules found in retinoblastomas and other tumors to bind E2F, either when assayed for binding to E2F activity in cell lysates, or when assayed for binding specifically to the E2F-1 protein (BANDARA et al. 1991; CHELLAPPAN et al. 1991; KAELIN et al. 1991). pRB's viral oncoprotein-binding pocket domain is necessary but not sufficient for binding to E2F. E2F binding also requires a C-terminal pRB domain that is outside of this pocket region (HIEBERT 1993; QIAN et al. 1992, QIN et al. 1992), and which specifically involves pRB C-terminal residues 785–806 (WELCH and WANG 1995).

The DNA binding activity termed E2F actually subsumes a family of heterodimeric transcription factors in which each heterodimer consists of one "E2F" and one "DP" polypeptide (see Slansky and Farnham, this volume, for review). Understanding of pRB–E2F interactions improved with the cloning and analysis of genes that encode the various E2F and DP polypeptides.

E2F-1, -2, and -3 share a conserved C-terminal pRB-binding domain that is embedded within a transcriptional activation domain (HELIN et al. 1992; KAELIN et al. 1992; SHAN et al. 1992; IVEY-HOYLE et al. 1993; LEES et al. 1993). In E2F-1, this C-terminal region contains an 18 amino acid segment that is both necessary and sufficient for pRB binding (HELIN et al. 1992). Ten of these 18 residues are conserved in E2F-2 and -3, suggesting that this region is sufficient for binding of these E2Fs to pRB as well (LEES et al. 1993).

pRB's ability to bind E2F-1 is greatly enhanced, both in vitro and in vivo through the heterodimerization of E2F-1 with the DP-1 protein (HELIN et al. 1993b; KREK et al. 1993). This suggests that pRB ineracts with the E2F/DP heterodimers that seem to be the most abundant form of E2F in the cell. Interestingly, while E2F-1 on its own appears to bind (albeit poorly) directly to the pRB pocket and a C-terminal region, a weak but direct interaction that has been detected between pRB and DP-1 is entirely pocket independent (BANDARA et al. 1994). This suggests that the interaction between E2F-1 and a pRB region that includes the pocket may be stabilized by an interaction between DP-1 and a pRB element outside of the pocket. Similar stabilizing interactions might be expected to be conferred by other DP proteins such as DP-2. Since interactions among pRB and E2F family members probably always involve an E2F/DP heterodimer in vivo, for convenience the presence of the DP moiety is implied but not specified further in this review.

The discovery of pRB:E2F complexes relied in large part upon the hypothesis that E1A dissociates such complexes through binding to pRB (BAGCHI et al. 1990). Recent reports now provide a mechanism through which E1A might accomplish this dissociation. Apparently, the LxCxE motif in the CR2 region of E1A binds directly to a part of pRB that is unoccupied in pRB:E2F complexes. Meanwhile, the CR1 region of E1A competes directly with E2F for binding to some other pRB site and blocks E2F from rebinding that site upon the equilibrium dissociation of E2F from pRB (FATTAEY et al. 1993a; IKEDA and NEVINS 1993; WU et al. 1993; BARBEAU et al. 1994). This topic is treated in detail elsewhere in this volume (Nevins, this volume).

2.3 Cell Cycle Dependent Regulation of pRB:E2F Complexes

Whereas studies of DNA tumor virus oncoproteins pointed the way towards the discovery of pRB:E2F complexes, studies of cell cycle control proteins have provided insight into how pRB:E2F complexes are regulated. Shortly after the identification of the RB tumor suppressor gene, pRB was found to be phosphorylated in a cell cycle dependent manner, becoming hyperphosphorylated as cells pass from the G_1 into the S phase of the cell cycle and subsequently becoming dephosphorylated to a basal hypophosphorylated state upon leaving mitosis (BUCHKOVICH et al. 1989; CHEN et al. 1989; MIHARA et al. 1989; LUDLOW et al. 1990; DECAPRIO et al. 1989, 1992).

The timing of pRB hyperphosphorylation closely coincides with passage of the cell through the G_1/S restriction point, which is when a cell commits itself to a round of DNA replication and cell division. This tight correspondence suggests that pRB hyperphosphorylation is a critical event in the cell's committing itself to completion of a cell cycle. pRB is phosphorylated at amino acid residues that fall within consensus target sites for the cdk family (LEES et al. 1991; LIN et al. 1991), and abundant evidence suggests that pRB is phosphorylated by at least one and possibly a series of these kinases (REED 1992). Currently, the best pRB kinase candidates are cyclin D/cdk4 and cyclin E/cdk2 (EWEN et al. 1993; KATO et al. 1993; HATAKEYAMA et al. 1994; MITTNACHT et al. 1994).

The hypophosphorylated pRB present in G_0 and G_1 cells is believed to be the biochemically active pRB form, largely because only hypophosphorylated pRB is bound with high affinity by the adenovirus E1A, SV40 T antigen, and human papilloma virus (HPV) E7 proteins during the course of viral oncoprotein-induced transformation (LUDLOW et al. 1989; IMAI et al. 1991; TEMPLETON et al. 1991). Importantly, E2F also binds only to the hypophosphorylated form of pRB (CHELLAPPAN et al. 1991; HELIN et al. 1992; KAELIN et al. 1992; SHAN et al. 1992), and free E2F can be released from pRB:E2F complexes upon pRB phosphorylation by the cyclin E/cdk2 or the cyclin A/cdk2 kinases in vitro (DYNLACHT et al. 1994b). Together, these observations suggest that hyperphosphorylation of pRB by cyclin-dependent kinases releases free E2F from pRB:E2F complexes, just as cells commit themselves to entrance into the late G₁ and S phases of the cell cycle (see lower portion of Fig.1).

pRB's ability to bind E2F-1 may be affected not only by pRB phosphorylation but also by E2F-1 phosphorylation. Phosphorylation of E2F-1 on serine residues 332 and 337 is reported to prevent the interaction of E2F-1 with pRB (FAGAN et al. 1994). These serine residues fall within cdk consensus phosphorylation sites and are phosphorylated by various cyclin/cdk species in vitro. In addition, these sites appear to be phosphorylated in the cell at about the same time that pRB is hyperphosphorylated and inactivated by cyclin/cdks (FAGAN et al. 1994). Together, these results suggest that the cell may disrupt or prevent formation of pRB:E2F complexes through phosphorylation of either the pRB or the E2F component or both.

As described below and elsewhere in this volume, E2Fs appear to promote cell cycle progression through their activation of growth-promoting

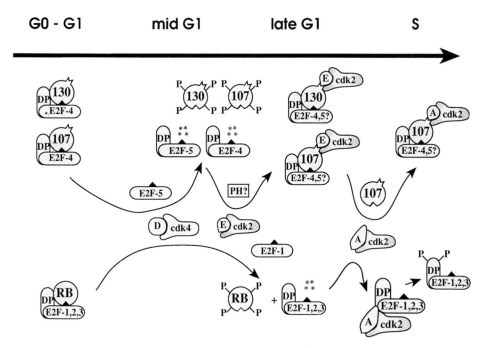

Fig.1. Cell cycle-dependent changes in E2F complexes. *Below,* pathways for pRB–E2F interactions; *above,* pathways for p107-E2F and p130–E2F interactions; *middle,* components whose synthesis or activation affects E2F complexes. PH is a hypothetical phosphatase. `****`, Active E2F. For further explanations please refer to the text

genes, while the binding of pRB to E2F represses the expression of such genes. Thus, the phosphorylation of pRB by cdks represents a critical link between the central cell cycle control machinery and the expression of E2F-responsive genes involved in cell cycle progression. However, pRB is not the only such link, since E2Fs are also bound and apparently regulated by two other members of the pRB family.

3 Interactions of E2Fs with pRB-Related Proteins

3.1 p107:E2F:Cyclin/cdk2 Complexes in Late G$_1$ and S Phases

At the time when pRB:E2F complexes were first detected, it was also apparent through gel shift experiments that some E2F multiprotein complexes do not contain pRB. The first of such complexes to be identified forms as cells enter the S phase of the cell cycle. Its components were found through the use of specific antibodies to include the pRB-related p107 protein and the cyclin A/cdk2 kinase (Mudryj et al. 1991; Cao et al. 1992; Devoto et al. 1992; Pagano et al. 1992a; Shirodkar et al. 1992; see S phase complex in Fig. 1).

The p107:E2F:cyclin A/cdk2 complexes are dissociated by the E1A onco-protein in the same way as the pRB:E2F complexes are dissociated by E1A (Mudryj et al. 1991; Dyson et al. 1992; Shirodkar et al. 1992). This disruption by E1A is hardly surprising, given that p107 is a prominent E1A binding protein that is highly related to pRB in its oncoprotein-binding domain (Yee and Branton 1985; Harlow et al. 1986; Ewen et al. 1991). It is somewhat surprising, however, that the HPV E7 oncoprotein, which binds and dissociates pRB:E2F complexes (Chellappan et al. 1992), binds but does not dissociate p107:E2F:cyclin A/cdk2 complexes (Arroyo et al. 1993; Lam et al. 1994). The basis for this difference is currently unclear.

The S phase specific p107:E2F:cyclin A/cdk2 complex has histone H$_1$ kinase activity, due to the cyclin A/cdk2 moiety (Mudryj et al. 1991; Cao et al. 1992; Devoto et al. 1992; Pagano et al. 1992a; Shirodkar et al. 1992). A very similar four-part complex, but containing cyclin E instead of cyclin A, appears in late G$_1$ cells and dissipates as the p107:E2F:cyclin A/cdk2 complex arises in S (Lees et al. 1992). This temporally distinct association of cyclin E/cdk2 and cyclin A/cdk2 with p107:E2F reflects the patterns of expression and kinase activities of cyclins E and A (Hunter and Pines 1991; Lew et al. 1991; Dulic et al. 1992).

It is important to recognize that p107 can bind to cyclin A/cdk2 or cyclin E/cdk2 in the absence of E2F (Faha et al. 1993; Peeper et al. 1993) and, similarly, that p107 and E2Fs can bind one another in the absence of a cyclin/cdk2 (Cobrinik et al. 1993). p107 appears to bind to cyclin/cdk2 through a "spacer" domain situated between the A and B segments of the p107 pocket

(EWEN et al. 1991, 1992; FAHA et al. 1992). This spacer is not required for E2F binding, while the p107 E2F-binding region is dispensable for cyclin/cdk2 binding (ZHU et al. 1993).

Taken together, the above observations indicate that p107:E2F:cyclin/cdk2 complexes form as a consequence of p107's independent intractions with E2Fs and cyclin/cdk2. Interestingly, p107 can function as an inhibitor of cell growth either through E2F binding or through cyclin/cdk2 binding (SMITH and NEVINS 1995; ZHU et al. 1993, 1995), suggesting that both of these interactions may have functional significance. However, functions of p107 that require its simultaneous interaction with E2F and a cyclin/cdk2 have yet to be identified.

The role of the kinase moiety within p107:E2F:cyclin/cdk2 complexes has been the subject of much speculation. Since the binding of such complexes to E2F recognition sequences localizes a cyclin/cdk2 kinase to specific promoters, it has been proposed that these complexes regulate gene expression by phosphorylating components of the transcription apparatus in situ at such promoters. Alternatively, since p107 also brings the cyclin/cdk2 kinase into close proximity with E2F, cyclin/cdk2 binding to p107 might facilitate E2F phosphorylation and modify E2F function (CAO et al. 1992; DEVOTO et al. 1992; SHIRODKAR et al. 1992). Yet another possibility is that cyclin/cdk2 kinases exert some function through their binding and phosphorylation of p107 itself (PEEPER et al. 1993). In fact, evidence supporting these or other roles of the p107:E2F-associated kinase is yet to be obtained.

Recent work suggests that p107 binding to cyclin/cdk2 inhibits cell growth (ZHU et al. 1995), and it is tempting to speculate that this may be due to an inhibition of cyclin/cdk2 kinase activity. If this is the case, however, the detected kinase activity of p107:E2F:cyclin/cdk2 complexes would reflect an incomplete inhibition in vitro. Thus far, the real function served by p107 binding to cylin/ckd2 remains an enigma.

3.2 p130:E2F and p107:E2F Complexes in G_0 and G_1 Phases

The existence of late G_1 and S phase p107:E2F:cyclin/cdk2 complexes suggested a cell-cycle dependence in the composition of E2F complexes, with pRB associating with and presumably regulating the activity of E2F in G_0 and early to middle G_1, and p107 associating with and regulating E2F in late G_1 and S. However, analyses of G_0 cells indicates that a protein other than pRB and p107 is often the major E2F-associated species.

Neither pRB nor p107 were detected in the predominant E2F complexes in resting human T cells or mouse fibroblasts (CHITTENDEN et al. 1993; COBRINIK et al. 1993). Nonetheless, in each of these cell types, the G_0 E2F complexes could be disrupted by E1A or other viral oncoproteins (MUDRYJ et al. 1991; PAGANO et al. 1992b; CHITTENDEN et al. 1993; COBRINIK et al. 1993). These observations suggested that E2F associates with a novel pRB-related protein initially termed "X" in G_0 cells (LEES et al. 1992; SHIRODKAR et al. 1992; CHITTENDEN et al. 1993), and that E1A displaces E2F from X in the same way that it displaces E2F from pRB and p107. Based upon this model, an affinity enrichment procedure was developed to detect proteins in G_0 fibroblast lysates that bind to both E2F and E1A. This approach led to the identification of the p130 protein as the major E2F-associated species in G_0 cells (COBRINIK et al. 1993).

p130 is a pRB-related protein (HANNON et al. 1993; LI et al. 1993) that shares with other pRB family members the capacity to be bound by the E1A, T antigen, and E7 viral oncoproteins (DYSON et al. 1992). However, p130 is far more closely related to p107 than it is to pRB. Due to this similarity antibodies against p130 and p107 often cross-react, and this necessitates caution in deducing the components of gel shift complexes through their antibody reactivities.

The affinity enrichment of proteins that bind to both E2F and E1A is a sensitive and specific way to assess the levels of pRB, p107, and p130 that are associated with E2F in cell lysates. With this approach p130:E2F complexes were detected in a variety of cell types that had been brought to quiescence by serum starvation including fibroblasts, keratinocytes, retinoblasts, and fetal kidney cells. However, pRB:E2F and p107:E2F complexes were also prominent in most of these G_0 cell extracts (COBRINIK et al. 1993). These observations support the view obtained through gel shift analyses that distinct kinds of E2F complexes coexist in G_0 and early G_1 cells (CHITTENDEN et al. 1993; see G_0–G_1 portion of Fig. 1).

The relative levels of E2F-associated pRB family members can vary among closely related cell types. Indeed, p107:E2F complexes appear to be more abundant than p130:E2F complexes in serum-starved mouse NIH 3T3 fibroblasts (LAM et al. 1994), whereas the converse is true in mouse BALB/c 3T3 fibroblasts or in primary fibroblasts (COBRINIK et al. 1993). Importantly, p130 becomes the most abundant E2F-associated protein in BALB/c 3T3 cells only after serum-starvation (Cobrinik, unpublished observation). Thus, association of E2F predominantly with p107 in serum-starved NIH 3T3 fibroblasts may reflect the relatively vigorous growth of NIH 3T3 cells and their inability to enter a G_0 state identical to that of BALB/c 3T3 or primary fibroblasts.

3.3 Binding of p130:E2F and p107:E2F to Cyclin/cdk2

An important structural distinction between the different pRB family members is that both p107 and p130 contain a spacer region within their pocket domains that is entirely absent from pRB (Ewen et al. 1991; HANNON et al. 1993; LI et al. 1993). For p107 this spacer segment directs stable binding to the cyclin A/cdk2 kinase (EWEN et al. 1991, 1992; FAHA et al. 1992; ZHU et al. 1993) and is presumed to direct binding to the cyclin E/cdk2 kinase as well. Since p130 associates with the cyclin A/cdk2 and cyclin E/cdk2 kinases in a manner analogous to p107 (HERRMANN et al. 1991; COBRINIK et al. 1993; FAHA et al. 1993; HANNON et al. 1993; LI et al. 1993), the p130 spacer is similarly implicated in directing cyclin/cdk2 binding.

As is the case for p107, p130 interacts simultaneously with E2F and cyclin/cdk2s. In vitro reconstitution experiments demonstrate the ability of p130:E2F complexes to associate with cyclin E/cdk2 and cyclin A/cdk2, and p130:E2F:cyclin E/cdk2 complexes have been detected in fibroblasts in the late G_1 portion of the cell cycle (COBRINIK et al. 1993).

Reconstitution studies further indicate that p130 and p107 are unable to bind stably either to cyclins E or A or to cdk2 alone; stable association requires both the cyclin and the cdk2 components (COBRINIK et al. 1993; PEEPER et al. 1993; SCHWARZ et al. 1993). This may explain the absence of cdk2 in p130:E2F or p107:E2F complexes in G_0 cells, despite cdk2 expression (but not cyclin E or A expression) in such cells. Furthermore, cyclin E/cdk2 and cyclin A/cdk2 are the only cyclin/cdk combinations yet found to be reconstituted into p107:E2F and p130:E2F complexes. Recent reports, however, suggest that cyclin D may be present in E2F complexes of quite distinct architecture (KIYOKAWA et al. 1994; AFSHARI et al. 1995; see Sect. 8.2 below).

Although reconstitution studies indicate that p130:E2F complexes can interact with cyclin A/cdk2 as well as with cyclin E/cdk2, p130:E2F:cyclin A/cdk2 complexes have not been identified in cell lysates. This may reflect the fact that p107 protein levels are dramatically induced near the G_1/S border (COBRINIK et al. 1993), and this would allow p107 to out-compete p130 for a position in the majority of such complexes. Taken together, the bulk of the evidence is consistent with the notion that formation of the various four part E2F complexes is determined largely by the availability of each of the component parts.

3.4 The Transition from G_0/G_1 p130:E2F and p107:E2F Complexes to S Phase p107:E2F:Cyclin A/cdk2 Complexes

Binding of pRB family members to E2F transcription factors appears to inhibit E2F-dependent transactivation of growth-related genes (see below). Thus, free, active E2F might be expected to be released from inhibitory interactions with p130 of p107 to allow the expression of such genes in late G_1. Indeed, in synchronized mouse fibroblasts and human T cells – where most E2F is bound to p130 or p107 in G_0 or early G_1 – free E2F increases during progression into middle to late G_1 (MUDRYJ et al. 1991; CHITTENDEN et al. 1993). Recent results suggest that at least some of this increase in free E2F is due to the release of E2F from p107 and p130.

p107 and p130 have recently been found to be phosphorylated in middle to late G_1 fibroblasts, concurrent with the induction of cyclin D gene expression but prior to the induction of cyclin E. Furthermore, ectopic expression of cyclin D/cdk4 – but not cyclin E/cdk2 – causes p107 to become hyperphosphorylated and to release free E2F (BEIJERSBERGEN et al. 1995). Together, these observations suggest that phosphorylation of p107 and perhaps p130 by cyclin D/cdk4 causes the release of free E2F in middle to late G_1 (see Fig. 1). Additional free E2F in middle to late G_1 cells may also result from the dramatic inductions of the E2F-5 and E2F-1 genes in these portions of the cell cycle (KAELIN et al. 1992; SHAN et al. 1992; SLANSKY et al. 1993; SARDET et al. 1995).

Earlier studies had indicated that p130 exists in two electrophoretically distinguishable phosphylation states that are each able to complex with E2F, as well as with E1A (DYSON et al. 1992; COBRINIK et al. 1993). Based upon the recent results of the Bernards group, however, the putatively hyperphosphorylated p130 detected in E2F complexes may in fact lack the phosphorylation of key residues that would abrogate E2F binding in middle to late G_1. Clearly, more detailed analyses are required to reveal the phosphorylation control of p130:E2F complexes.

The insight that p107 and possibly p130 release E2F in middle to late G_1 raises the question of how these pRB-related proteins come to be complexed with E2F (along with cyclin/cdk2 kinases) in late G_1 and S phase cells. One possibility is that p107 and p130 are dephosphorylated following the induction of cyclin E, which would be expected to restore p107's and p130's E2F binding capabilities. An additional but not mutually exclusive possibility is that late G_1 and S phase E2F complexes contain newly synthesized p107 that has never been exposed to the cyclin D/cdk4 kinase. This would provide a rationale for the dramatic induction of p107 detected near the G_1/S border (COBRINIK et al. 1993). A critical feature for both of these explanations is that the cyclin D/cdk4 kinase is incapable of phosphorylating p107 or p130 in late G_1 or S. In fact, growing evidence suggests that this kinase is indeed inactivated in late G_1 (reviewed in WEINBERG 1995).

4 Specific Interactions Between E2Fs and pRB Family Members

The finding that E2Fs interact simultaneously with multiple pRB family members raised the possibility that the different pRB family members complex with biochemically and functionally distinct E2Fs. Recent studies reveal that the different E2F species (in association with DP family members) do indeed interact with distinct pRB family members in mammalian cells.

4.1 Biochemically Distinct Proteins Interact with p107 and pRB

The first hints of specific interactions between E2F and pRB family members were suggested by the release of apparently distinct "free" E2F gel shift species upon disruption of different E2F multiprotein complexes with E1A (CHITTENDEN et al. 1993). Meanwhile, ^{32}P-labeled E2Fs that coimmunoprecipitate with pRB and p107 were found to have distinct electrophoretic mobilities, and the E2Fs that coimmunoprecipitate with pRB were recognized by a certain E2F-1 antibody, while those that coimmunoprecipitate with p107 were not (DYSON et al. 1993). Similar coimmunoprecipitation assays revealed interactions between ^{32}P-labeled E2F-1,-2,-3 with pRB but not with p107 (LEES et al. 1993).

4.2 Selective Interactions of E2F-1 and E2F-4

That the pRB- and p107-bound E2Fs are actually distinct gene products is now evident from the recent cloning of a novel E2F family member, E2F-4, that preferentially associates with p107 in vivo (GINSBERG et al. 1994; BEIJERSBERGEN et al. 1994). In ^{32}P-labeled ML-1 cells E2F-4 is associated with p107 but not with pRB, while conversely E2F-1 is associated with pRB but not with p107 (BEIJERSBERGEN et al. 1994). The same specific interactions are also apparent in ^{35}S-labeled U937 cells, except that in these cells low levels of E2F-4 are also bound to pRB (GINSBERG et al. 1994). This suggests that there is a preferential but not absolute targeting of E2F subtypes by pRB and p107.

While E2F-4 clearly has a higher affinity for p107 than for pRB, its ability to distinguish between p107 and the more closely related p130 protein is unclear. Indeed, E2F-4 might associate with either p107 or p130, depending upon which is present at higher levels in the cell (VAIRO et al. 1995). It may also be relevant that E2F-4 exists as multiple differentially phosphorylated species (BEIJERSBERGEN et al. 1994; GINSBERG et al. 1994), and these different forms might exhibit different affinities for p107, p130, and perhaps pRB as well.

In contrast to E2F-4, an additional E2F family member, E2F-5, seems to interact specifically with p130 in lysates of untransfected cells (R. Bernards, personal communication). Whether this specific targeting of E2F-5 reflects its higher affinity for p130 than for p107, or the expression of E2F-5 in a confined cell cycle compartment (SARDET et al. 1995) where p130 levels exceed p107 levels, is yet to be determined.

The p107–E2F-4 interaction shares features with the more extensively characterized pRB–E2F-1 interaction. In particular, heterodimerization between E2F-4 and a DP protein such as DP-1 appears to be essential for binding of p107 to E2F-4 (BEIJERSBERGEN et al. 1994). Furthermore, E2F-4/DP-1 complexes bind to the same E2F DNA recognition sequence that is recognized by E2F-1/DP-1 (BEIJERSBERGEN et al. 1994; GINSBERG et al. 1994), and this binding specificity is expected to be retained in p107:E2F-4 complexes. Since E2F-4 is expressed throughout the cell cycle (GINSBERG et al. 1994; SARDET et al. 1995), E2F-4 is an excellent candidate for the E2F species present in p107: E2F:cyclin/cdk2 complexes in late G_1 and S phase cells, but this has not yet been demonstrated directly.

The specific interactions between E2F and pRB family members, which are evident in lysates of unmanipulated cells, are generally compromised in vitro. For example, in contrast to the preferential binding of E2F-1 to pRB versus p107 in unmanipulated cell lysates, in vitro translated E2F-1 appears to bind to GST-p107 and GST-pRB fusion proteins with similar affinities (CRESS et al. 1993). The specificity of these interactions may also be compromised when the corresponding proteins are ectopically expressed at high levels in vivo, such as in a yeast two-hybrid assay (where E2F-1 interacts equally productively with pRB, p107, and p130) or in transfected cells (SARDET et al. 1995; R. Bernards, personal communication). The ability of p107 to bind E2F-1 under these circumstances highlights the possibility that p107:E2F-1 or other low affinity complexes might indeed form and have functional consequences in vivo, particularly in situations when a preferred E2F or pRB family binding partner is absent.

4.3 Specific E2F Multiprotein Complexes in *Xenopus* and *Drosophila*

While E2F multiprotein complexes have been best characterized in mammalian cells, such complexes have also been detected in cells of nonmammalian species including *Xenopus* and *Drosophila*. In cultured *Xenopus* cells derived from metamorphosing frog, E2F associates with pRB as well as with cdk2 (PHILPOTT and FRIEND 1994). In analogy with the situation in mammalian cells, E2F seems likely to associate with cdk2 through a complex that contains *Xenopus* p107 or p130 and cyclin E or A homologues. Yet other evidence of antigenically distinct forms of E2F in *Xenopus* oocytes supports the notion that *Xenopus* p107 and pRB homologues interact with distinct E2F species in a manner reminiscent of E2F control in mammalian cells.

In *Drosophila* a rather different situation may apply. Thus far one and only one *Drosophila* homologue of each of the E2F, DP, and pRB gene family members has been identified, and low stringency probing of Southern blots suggests the absence of closely related genes (DYNLACHT et al. 1994a; OHTANI and NEVINS 1994; and N. Dyson, personal communication). Moreover, only one free E2F species and one pRB:E2F complex have been identified in gel shift analyses, and these complexes react completely with antibodies directed against *Drosophila* E2F, DP, and pRB. These findings suggest that only single E2F, DP, and pRB-like proteins are involved in the control of E2F-regulated genes in *Drosophila*, and they raise the possibility that p107:E2F and p130:E2F complexes perform a function that is unique to vertebrates.

5 Transcriptional Effects of E2F Multiprotein Complexes

In order to identify the cell growth control functions that are performed by the different E2F multiprotein complexes it is critical to define at the biochemical level the functional consequences of these interactions. Thus far the major direct consequences of interactions between E2F and pRB family members lay in the transcriptional control of cell growth-related genes.

5.1 Inhibition of E2F-Dependent Transcription by pRB Family Members

The binding of pRB to a segment within the E2F-1 transactivation domain suggests that pRB antagonizes E2F-1 function by blocking the interactions through which E2F-1 stimulates the basal transcription apparatus (HELIN et al. 1992; KAELIN et al. 1992). Moreover, the strong conservation of the transactivation domain among all E2Fs (IVEY-HOYLE et al. 1993; LEES et al. 1993; BEIJERSBERGEN et al. 1994; GINSBERG et al. 1994; SARDET et al. 1995) suggests that each of the pRB family members might inhibit E2F-mediated transactivation through a similar mechanism. In fact, numerous studies indicate that pRB family members inhibit E2F-dependent transcription.

In transient transfection assays pRB has been found to inhibit synthetic E2F-dependent promoters, the viral E2 promoter, and the E2F-dependent promoters of the c-*myc* and *cdc-2* genes, while coexpression of viral oncoproteins abrogates these effects (DALTON 1992; HAMEL et al. 1992; HIEBERT et al. 1992; ZAMANIAN and LA THANGUE 1992). While being suggestive, these findings do not show conclusively that pRB directly regulates E2F, since overexpression of pRB family members might have pleiotropic effects on cell growth that affect E2F-dependent promoters indirectly. Indeed, some promoters for which there is currently no evidence of direct control by pRB are also repressed, possibly indirectly, by the ectopic expression of a transfected RB gene (for example, see ROBBINS et al. 1990).

To address these concerns, FLEMINGTON et al. (1993) and HELIN et al. (1993a) constructed fusion genes encoding the DNA binding domain of the yeast Gal4 transcription factor fused to the E2F-1 transactivation domain. These Gal4–E2F-1 fusion proteins stimulated transcription from promoters containing Gal4 DNA binding sites, and this transactivation was inhibited by pRB. Importantly, transactivation by similar fusion proteins, but having mutations that block pRB binding, was insensitive to pRB overexpression. Thus, in transfection experiments pRB inhibits E2F-1 dependent transactivation through direct protein–protein interactions.

The pRB-related p107 and p130 proteins also inhibit E2F-dependent transcription in transfection experiments (SCHWARZ et al. 1993; ZAMANIAN and LA THANGUE 1993; ZHU et al. 1993; Cobrinik, unpublished observations). As might be expected from their known abilities to bind E2F-4, p107 and p130 inhibit ectopically expressed E2F-4 in such experiments as well (BEIJERSBERGEN et al. 1994; GINSBERG et al. 1994; VAIRO et al. 1995). It seems likely that these pRB-related proteins inhibit transcription through direct contacts with their cognate E2Fs, in a manner similar to that which was demonstrated directly for pRB and E2F-1. Perhaps anomalously, pRB also antagonizes E2F-4, while p107 and p130 antagonize E2F-1 in transfection experiments. These results attest to the lack of specificity that may occur in transfection and overexpression assays.

5.2 Repression of E2F-Regulated Genes by pRB

While the above analyses show that pRB antagonizes E2F-dependent trans-activation, pRB:E2F complexes may have a more significant role as repressors of genes that contain E2F DNA binding sites. In transfections of a series of pRB$^{(+)}$ cells, insertion of E2F recognition sites into otherwise strong promoters represses promoter activity, while coexpression of E1A converts these E2F sites from negative to positive transcriptional elements (WEINTRAUB et al. 1992). Assuming that E1A is working through binding of pRB family members, these findings indicate that pRB or related proteins actively repress promoters that contain E2F binding sites.

Importantly, in three cell lines that lack functional pRB (SAOS-2, C33A, and HTB-9), insertion of E2F binding sites into a strong promoter increases promoter activity (WEINTRAUB et al. 1992). Since at least p107 is present in SAOS-2 and C33A cells (ZHU et al. 1993; Cobrinik, unpublished data), this observation suggests that p107 and p130 fail to repress E2F-dependent transcription when expressed at endogenous levels in these cells. While this finding is consistent with the possibility that pRB but not p107 or p130 is capable of promoter repression, the abilities of p107 or p130 to repress transcription via distinct E2F recognition sequences or in other cell types is yet to be established. In this regard it may be significant that E7 stimulates the E2F-dependent adenovirus E2 promoter in several pRB$^{(-)}$ breast tumor cell lines (CARLOTTI and CRAWFORD 1993), and might do so throught the release of p107- or p130-mediated repression.

While a role of E2F sites in promoter repression was first established with heterologous promoters as described above, repression has also been detected in a series of cellular promoters. For example, removal of a fragment containing E2F sites derepresses the cdc2 promoter in quiescent cells (DALTON 1992). Similarly a G_0/G_1-specific repression is attributable specifically to E2F recognition sequences in the B-myb and E2F1 genes (LAM and WATSON 1993; HSIAO et al. 1994; JOHNSON et al. 1994; NEUMAN et al. 1994) and repressor activity may be attributable to the E2F sites in the c-myc promoter as well (N. Hay, personal communication). In these cases the primary role of the E2F binding sites appears to be in maintaining a low level of gene expression in the G_0 and G_1 phases of the cell cycle, rather than in stimulating gene expression in late G_1 or S.

Recent evidence indicates that promoter repression requires pRBs presence at the promoter. QIN et al. (1995) have found that a mutant of E2F-1 that is incapable of both transactivation and pRB binding can nevertheless overcome promoter repression by pRB. Presumably, this mutant E2F-1 binds to promoter E2F sites and thereby prevents pRB from binding such sites via cellular E2Fs.

In addition, Dean and colleagues have found that a Gal4-pRB fusion protein can repress transcription when localized to promoter Gal4 binding sites in the same way that pRB represses transcription when localized to E2F sites. Furthermore, the Dean group has found that promoter-localized pRB represses transactivators for which pRB has demonstrable – albeit relatively low – affinity, such as the c-*myc*, Elf-1, and PU-1 proteins (RUSTGI et al. 1991; HAGEMEIER et al. 1993a; WANG et al. 1993), but pRB fails to repress transactivators such as VP-16 or SP-1 for which it has no detectable affinity (WEINTRAUB et al. 1995). One interpretation of these results is that pRB is localized to certain promoters through its high-affinity binding to E2F and that pRB represses such promoters through independent lower affinity interactions with other transactivators.

The ability of pRB to repress entire promoters provides a powerful mechanism for the cell cycle specific regulation of gene expression. Through promoter repression pRB not only directly controls the activity of E2F, but pRB also in effect controls a host of other transcription factors that might not be cell cycle regulated in other promoter contexts.

5.3 Transcription Control by Specific E2F and pRB Family Members

A vexing problem in understanding the control of E2F-dependent genes is the identification of the specific E2F and pRB family members that regulate specific promoters. In fact, it is currently unclear whether absolute specificity of this kind exists. In one report overexpression of E2F-1 was able to stimulate some but not other E2F-regulated promoters, perhaps suggesting that E2F-1 targets some promoters while distinct E2Fs target others (LI et al. 1994). However, since overexpression of E2Fs or pRB family members can lead to non-physiologic interactions (as discussed above), insight into the roles of distinct E2F complexes in the transcription of selected genes might be more clearly obtained by ablating the functions of such proteins from cells.

One way to ablate the function of specific pRB family members is through the expression of mutant DNA tumor virus oncoproteins that bind to a restricted group of such proteins. Indeed, a mutant E7 protein that binds with high affinity to p107, but poorly if at all to pRB, has been found to derepress the B-*myb* gene in mouse NIH 3T3 fibroblasts (LAM et al. 1994). While this suggests that B-*myb* repression is mediated by p107, it is also possible that the mutant E7 binds pRB sufficiently well in vivo for it to release pRB-mediated repression. Should it turn out that B-*myb* is indeed specifically repressed by p107:E2F complexes in these cells, it will be of interest to determine whether similar regulation applies to other genes or, indeed, to the B-*myb* gene in other cell types.

A second means of ablating the function of specific pRB or E2F family members is through production of cells in which one or another of these genes has been knocked out through gene targeting. Indeed, intriguing results have already been obtained in analyses of Rb$^{(-/-)}$ mouse embryo fibroblasts (MEFs). Although wild-type MEFs have almost undetectable levels of pRB:E2F complexes (COBRINIK et al. 1993), Rb$^{(-/-)}$ MEFs exhibit a specific derepression of the cyclin E gene during serum starvation (HERRERA et al. 1995). Ongoing work is directed towards determining whether pRB normally represses the cyclin E promoter in MEFs via E2F recognition sequences (Y. Geng, and R.A. Weinberg, in preparation).

6 When the Job Is Done: S Phase Shut-Off of E2F-1, -2, and -3 by Cyclin A/cdk2

The way in which phosphorylation of pRB and pRB-related proteins frees E2Fs from inhibitory multiprotein complexes as cells approach the G_1/S border is described above. However, this understanding leads to yet another puzzle: how does the released, free E2F become inactivated at the time when most E2F-dependent genes are being shut off after the cell has entered S phase?

As discussed in Sect. 3.4, E2F-4 and -5 may be brought under the control of newly dephosphorylated or newly synthesized p107 in S phase, although concrete evidence for this is yet to be obtained. In contrast, E2F-1, -2, and -3 cannot bind to their pRB regulator in S phase, since pRB remains hyper-phosphorylated throughout this portion of the cell cycle. Instead, E2F-1, -2, and -3 may be inactivated in S phase through their binding to the cyclin A/cdk2 kinase and the subsequent phosphorylation of their associated DP proteins (DYNLACHT et al. 1994b; KREK et al. 1994; XU et al. 1994).

6.1 Association of E2F-1 with Cyclin A/cdk2

Cyclin A is a major E2F-1 binding protein in cell lysates (KREK et al. 1994), and cyclin A, but not cyclins E or B, binds to E2F-1 in vivo as well as in reconstitution assays in vitro (DYNLACHT et al. 1994b; KREK et al. 1994; XU et al. 1994). The association of cyclin A with E2F-1 depends upon the presence of cdk2 (KREK et al. 1994; XU et al. 1994), just as cdk2 is required for cyclin A to bind to p107:E2F and p130:E2F complexes (as discussed in Sect. 3.3). However, the direct binding of cyclin A/cdk2 to E2F-1 differs from its indirect association with E2Fs, through p107, in the prominent S phase E2F complexes described earlier (see Sect. 3.1). Indeed, interaction of cyclin A/cdk2 with E2F-1 involves E2F-1 N-terminal residues, and not the E2F-1 C-terminus used for pRB binding (KREK et al. 1994; XU et al. 1994).

6.2 Phosphorylation of DP-1 by Cyclin A/cdk2

In vitro, cyclin A/cdk2 can phosphorylate both DP-1 and E2F-1, and these phosphorylations do not depend upon E2F-1/DP-1 heterodimerization (DYNLACHT et al. 1994b; XU et al. 1994). In vivo, however, DP-1 is phosphorylated by cyclin A/cdk2 only after E2F-1/DP-1 heterodimers form, since DP-1 is not phosphorylated when bound to a truncated, noncyclin A-binding E2F-1 (KREK et al. 1994). While E2F-1 is also phosphorylated in vivo, this phosphorylation is independent of its ability to bind cyclin A/cdk2 and is not thought to be relevant to the inactivation of E2F-1 in S phase (KREK et al. 1994).

In vitro, cyclin A/cdk2 phosphorylation of DP-1 abrogates DNA binding by E2F-1/DP-1 heterodimers and inhibits E2F-1 dependent transactivation, while phosphorylation of E2F-1 has no such effects (DYNLACHT et al. 1994b; XU et al. 1994). These same regulatory events may also occur in vivo. Indeed, DP-1 is phosphorylated in S phase, coincident with the activation of the cyclin A/cdk2 kinase. Furthermore, an apparently free E2F-1 gel shift complex detected in late G_1 cells appears to lose its DNA binding capability during S phase, whereas this E2F-1 complex persists through S and G_2 in cells that express a mutant E2F-1 that cannot bind to cyclin A.

The above findings suggest that E2F-1 acts as an intermediary in directing the cyclin A kinase to phosphorylate DP-1 and cause E2F-1/DP-1 heterodimers to be released from their promoter DNA binding sites in S phase cells (see Fig. 1). This interpretation is buttressed by transfection experiments showing that cyclin A inhibits E2F-1-dependent transcription in a fashion that requires the E2F-1 N-terminal cyclin A binding domain (KREK et al. 1994). Thus, the induction of cyclin A in S phase cells may block both DNA binding and transactivation by E2F-1.

E2F-2 and -3 are likely to be shut off in S phase in the same fashion as E2F-1. The E2F-1 N-terminal cyclin A-binding region is conserved in E2F-2 and E2F-3 (IVEY-HOYLE et al. 1993; LEES et al. 1993), and these E2Fs also stably associate with cyclin A/cdk2 (W. Krek and D.M. Livingston, personal communication). The N-terminal cyclin A binding region is absent from E2Fs-4 and -5 (BEIJERSBERGEN et al. 1994; GINSBERG et al. 1994; SARDET et al. 1995), suggesting that these E2F subtypes are not shut off in S phase through cyclin A/cdk2.

These observations add a key component to our understanding of the cell cycle dependent regulation of E2F-1, 2, and 3. As described above, these E2Fs are bound and inhibited by pRB in G_0 and G_1 cells, but are released from pRB through pRB and possibly their own phosphorylation at the hands of late G_1 kinases such as cyclin D/cdk4 and cyclin E/cdk2. At this point, these E2Fs are free to activate genes whose expression is critical to entrance into the S phase of the cell cycle. After cyclin A is synthesized at the start of S phase E2F-1, 2, and 3 lose their ability to bind DNA and stimulate gene expression due to the cyclin A/cdk2-dependent phosphorylation of DP-1 (see Fig. 1).

7 Inhibition of Cell Growth Through Interactions Among E2Fs and pRB Family Members

A role for pRB family members in cell growth inhibition was inferred from evidence that their association with DNA tumor virus oncoproteins contributes to cell transformation, as well as from the frequent mutational inactivation of pRB in human tumors (see Nevins, this volume, for review). Indeed, growth inhibition by pRB is readily manifested by the arrested cell growth or inhibited tumorigenesis caused by its ectopic expression in various pRB-negative as well as some pRB-positive cells (HUANG et al. 1990; GOODRICH et al. 1991; TEMPLETON et al. 1991; HINDS et al. 1992; FUNG et al. 1993). More recently, ectopic expression of p107 and p130 have also been found to inhibit cell growth (ZHU et al. 1993; CLAUDIO et al. 1994; SMITH and NEVINS 1995).

Evidence that E2F multiprotein complexes repress transcription of cell growth-related genes suggested that E2Fs may be critical targets through which pRB and related proteins inhibit cell growth. In the case of pRB this notion was reinforced by evidence that mutant forms of pRB that are nonfunctional in cell growth inhibition assays are also unable to bind E2Fs (HUANG et al. 1992; QIAN et al. 1992; QIN et al. 1992; HIEBERT 1993). However, this hardly proves that E2F is a critical pRB target, since pRB binds in vivo to a number of other proteins, and its interactions with these other targets are usually abrogated in nonfunctional pRB mutants as well (DOWDY et al. 1993; DURFEE et al. 1993; FATTAEY et al. 1993b; GU et al. 1993; QIAN et al. 1993; WANG et al. 1993; WELCH and WANG 1993).

7.1 Role of E2F-Dependent Promoter Repression in pRB-Induced Senescence

The case for E2F being a critical target for pRB-mediated growth suppression has only recently become compelling and at present applies only to a single assay system that is thought to reflect pRB-induced senescence.

Ectopically expressed pRB induces a growth arrest phenotype that resembles senescence in SAOS-2 cells (TEMPLETON et al. 1991) whereas ectopic expression of E2F-1 overcomes this phenotype (ZHU et al. 1993; QIN et al. 1995). Moreover, ectopic expression of an E2F-1 mutant that is incapable of transactivation and pRB binding but retains the ability to bind to the E2F recognition sequence also overcomes this pRB-induced senescence (QIN et al. 1995). The absence of the transactivation domain in this E2F-1 mutant ensures that pRB-induced senescence is not overcome through spurious transactivation of growth-related genes. Rather, this mutant must overcome pRB-induced senescence by binding to promoter E2F sites and preventing pRB from localizing to such sites via endogenous E2Fs.

Since the binding of pRB to promoter E2F sites is required for promoter repression (see Sect. 5.2), the functioning of this E2F-1 mutant suggests a role for promoter repression in pRB-induced senescence. Since E2F trans-activation is dispensable for overcoming pRB-induced senescence, gene promoters that are relevant to this process are apparently derepressed solely through the exclusion of pRB.

While blocking pRB's localization to promoter E2F sites is sufficient for overcoming pRB-induced senescence, it is insufficient to overcome pRB-mediated G_0/G_1 arrest as measured 48 h after transfection of the *RB* gene (QIN et al. 1995). Apparently, transactivation by E2F-1 or other dominant growth-promoting factors is required for overcoming this short term G_0/G_1 arrest imposed by ectopically expressed pRB.

7.2 Role of the pRB C Pocket in pRB-Induced Senescence

Recent evidence suggests that pRB-induced senescence involves an interaction involving a pRB C-terminal "C pocket" region in addition to the pRB–E2F interaction (WELCH and WANG 1995). Ectopic expression of this pRB C pocket blocks the induction of SAOS-2 flat cells by full length pRB in a dominant negative fashion but has no effect on pRBs binding to E2F. Along with earlier work, this observation suggests that pRB is localized to cellular promoters through its interaction with E2F and simultaneously interacts through its C pocket with some critical target molecule in the course of pRB-induced senescence.

The c-Abl protein may be a target of the pRB C pocket since the C pocket binds c-Abl and thereby inhibits c-Abl kinase activity (WELCH and WANG 1993). Furthermore, the pRB C pocket c-Abl interaction can occur simultaneously with the pRB:E2F interaction (WELCH and WANG 1995). However, it is currently unclear whether c-Abl or some other target mediates the critical pRB C pocket function that is required for pRB-induced senescence.

7.3 Growth Inhibition
by the pRB-Related p130 and p107 Proteins

The pRB-related p130 and p107 proteins inhibit cell growth upon their ectopic expression in various kinds of cells (ZHU et al. 1993; CLAUDIO et al. 1994; SMITH and NEVINS 1995). However, growth inhibition by p107 has so far been detected only in cells that are believed already to express normal p107 (ZHU et al. 1993; SMITH and NEVINS 1995). Growth inhibition by p130 has been demonstrated in potentially p130[-] HONE-1 cells, yet this is not known to be a property specific to p130 (CLAUDIO et al. 1994). Thus, it is unclear whether the growth inhibition caused by the ectopic expression of p107 or p130 reflects physiologically relevant properties of these proteins. Nonetheless, analysis of the interactions that mediate their growht inhibition may provide clues to the normal functioning of p107 and p130.

Most important for the current discussion of E2F multiprotein complexes is the observation that p107 mutants with impaired binding either to E2Fs or to cyclin/cdks are fully capable of cell growth inhibition while mutants with impaired binding to both E2F and cyclin/cdks lose this ability (SMITH and NEVINS 1995; ZHU et al. 1993). This suggests that p107 (and perhaps p130) can initiate independent growth-inhibitory pathways via E2F binding and via cyclin/cdk binding. Further work is required to establish the mechanism and the physiological relevance of these phenomena.

8 Role of E2F Multiprotein Complexes in Biological Processes

Changes in E2F multiprotein complexes occur not only within the context of cell cycle control but also within the context of the changing growth control regimens experienced by cells during organismal development, cellular senescence, and signal transduction. This section addresses our current understanding of the interactions among E2Fs and pRB family members in these processes.

8.1 Changes in E2F Multiprotein Complexes During Development

The first indications of developmental changes in E2F multiprotein complexes came from analyses of differentiating mouse embryonal carcinoma (EC) and embryonic stem (ES) cells. The absolute levels of E2F complexes have been found to be reduced during EC cell differentiation (LA THANGUE and RIGBY 1987; REICHEL et al. 1987), while the proportion of free E2F is decreased and that of E2F multiprotein complexes increased during both EC and ES cell differentiation (LA THANGUE et al. 1990; PARTRIDGE and LA THANGUE 1991; REICHEL 1992). While the presence of pRB in the complexes that form during differentiation has not yet been firmly established, it may be relevant that pRB is largely hyperphosphorylated and therefore unable to bind E2F in ES cells (possibly reflecting their extremely brief G_1 period). During ES cell differentiation, however, pRB becomes hypophosphorylated and capable of interacting with E2F (SAVATIER et al. 1994).

Changes in E2F complexes during ES cell differentiation are paralled by changes in E2F complexes during murine embryogenesis. Abundant free E2F was detected in 8.5-day embryos, and this free E2F species was gradually diminished, while E2F multiprotein complexes increased, by 17.5 days of gestation (PATRIDGE and LA THANGUE 1991). The exact components of the E2F complexes formed during embryogenesis are unclear, although the expression of pRB mRNA by 9.5 days of gestation (BERNARDS et al. 1989) suggests that pRB may be one such component.

Interestingly, analogous changes occur during *Xenopus* development. E2F appears to be largely free in oocytes and early embryos, while pRB:E2F complexes are first detected at low levels in the middle blastula phase (stage 12) and become prominent in embryo extracts only in late development (stages 35, 36) when the amount of total E2F activity is diminished (PHILPOTT and FRIEND 1994). Complexes containing cdk2 (presumably along with cyclins and a pRB-related protein) are not yet detectable at this late embryonic stage but are present in cells derived from metamorphosing frogs (PHILPOTT and FRIEND 1994).

In situ immunofluorescence analysis of late-stage mouse embryos indicates that the levels of pRB can vary considerably during the development of certain epithelial tissues. For example, pRB is abundantly expressed in differentiating keratinocytes and differentiating intestinal enterocytes but is undetectable in the relatively undifferentiated skin basal cells and colonic mucosa crypt cells (SZEKELY et al. 1992). To the extent that pRB levels reflect levels of pRB:E2F complexes, the cell growth inhibition mediated by such complexes would seem to increase during epithelial cell maturation.

8.2 Changes in E2F Multiprotein Complexes During Differentiation, Senescence, and Cytokine Signaling

While the various kinds of E2F complexes described throughout this review are the major ones detected in a wide range of growing cell types, additional E2F complexes have been detected recently in cells undergoing differentiation or senescence.

The first of these novel complexes to be detected contains E2F in association with cyclin D proteins. KIYOKAWA et al. (1994) found that both E2F activity and pRB coimmunoprecipitate with cyclin D_3 in extracts of HL60 erythroleukemia cells that were induced to differentiate, but neither E2F nor pRB coimmunoprecipitated with cyclin D_3 in extracts of untreated cells. From our current understanding of pRB's binding domains it appears that pRB is capable of binding simultaneously to cyclin D_3 and E2F. This is because D cyclins bind to pRB via an LxCxE motif reminiscent of that found in the adenovirus E1A CR2 domain (DOWDY et al. 1993). pRB can bind simultaneously to E1A LxCxE sequences and to E2F (see Sect. 2.3), and pRB may therefore be able to bind simultaneously to cyclin D LxCxE sequences and E2F in the same way. However, this is yet to be demonstrated.

Cyclin D has also been detected in E2F gel shift complexes in late-passage human primary cells undergoing senescence (AFSHARI et al. 1995). Intriguingly, these complexes also contain the cdk inhibitor p21 (also termed cip1, WAF1, Sdi1, CAP20, and PIC1) that appears to be induced in senescing cells. In toto, these complexes are believed to consist of p107:E2F:cyclin D/cdk2:p21. The architecture of such complexes could be similar to that proposed above for the pRB:E2F:cyclin D_3 complexes, but in this instance with cyclin D_3 binding to p107 (see EWEN et al. 1993), and associating independently with cdk2 and p21. To add a bit more complexity, AFSHARI et al. (1995) detect complexes consisting of pRB:E2F:cyclin E/cdk2:p21. An architecture for this complex cannot be inferred from our current understanding of the component proteins.

Additional E2F complexes containing the p130 protein have also been detected in differentiating melanoma cells. Interestingly, a p130 complex that may consist only of p130:E2F is detected following a reversible in vitro differentiation regimen, while a more slowly migrating p130-containing complex, perhaps also containing p21, is detected following irreversible differentiation (JIANG et al. 1995; S. Chellappan, H. Jiang, and P. Fisher, personal communication).

As with p107:E2F:cyclin A/cdk2 and related complexes described in Sect. 3, the physiological role of these novel multiprotein complexes is completely obscure. Perhaps of foremost importance in placing such complexes in an understandable framework will be determining whether they exert a specific biological function or form merely as a consequence of the high level of concurrent expression of the component proteins. With regard to this issue it may be relevant that p21 can stably interact with p107:E2F:cyclin D/cdk2 complexes but causes p107:E2F:cyclin A/cdk2 complexes to dissociate (AFSHARI et al. 1995). This suggests that p21 may play an active role in controlling the composition of certain E2F complexes.

A recent report suggests that E2F complexes may be regulated through cytokine signaling pathways. MELAMED et al. (1993) have found that the DNA binding activity of free E2F and E2F multiprotein complexes is rapidly reduced following treatment of Daudi cells with interferon-α or -β or with interleukin-6. This decrease in E2F activity precedes cytokine-induced suppression of c-*myc* expression and growth arrest, suggesting that it may play a part in these cytokine responses. Addition of EDTA to the extracts from cytokine-treated cells permits pRB:E2F gel shift complexes to be detected. Thus, pRB appears to act as an inhibitor of E2F DNA binding in the treated cell extracts, reminiscent of the in vitro E2F-inhibitor activity described in Sect. 2.1. It is currently unclear how this apparant "E2F-inhibitor" activity reflects cytokine signaling pathways.

9 Are Additional E2F Interactions Yet To Be Discovered?

The recent onslaught of novel E2F multiprotein complexes described here suggests that additional interactions involving E2Fs, pRB family members, cyclins, cdk's, and cdk inhibitors are yet to be identified and their biological functions deduced. Furthermore, investigating the mechanism of promoter repression by pRB:E2F complexes promises to reveal novel interactions that will have important implications for the general understanding of eukaryotic transcription. In addition, E2Fs interact in interesting ways with various viral proteins, and some of these interactions are discussed elsewhere in this volume (NEVINS). However, there are yet other kinds of E2F interactions that are waiting to be explored.

One group of E2F interactions yet to be investigated are those that are utilized when E2Fs stimulate transcription. One report has shown that E2F can bind to the TATA box binding protein (TBP) in vitro (HAGEMEIER et al. 1993b). Whether this interaction is physiologically relevant is unclear, however, since other proteins known to bind TBP in vitro probably interact instead with specific TBP-associated factors (TAFs) in the course of transcriptional activation (THUT et al. 1995). It would be interesting to learn whether E2Fs stimulate transcription through specific TAFs that are assembled within the basal transcription apparatus. Understanding these interactions may contribute to the development of therapeutic reagents that block E2F-dependent transcription in $Rb^{(-)}$ cells.

A second kind of E2F interation exists at this point only as a hypothetical possibility. All known E2F proteins contain a highly conserved region, termed the "marked box" (LEES et al. 1993; SARDET et al. 1995) that, at least in the case of E2F-1, is required for binding to the adenovirus E4 protein (HELIN and HARLOW 1994). The interaction of E2F with E4 stabilizes two E2F heterodimers bound to nearby sites on the E2 promoter and thereby contributes to a high level of E2 gene expression (see Nevins, this volume, for review). As the "marked box" is unlikely to have been conserved specifically to promote E4 binding upon a cell's infection with adenovirus, it seems plausible that a cellular protein binds to this region and perhaps performs a function analogous to that of E4 in stabilizing E2F complexes at selected cellular promoters.

10 Summary and Prospects

This review has described a complex and expanding network of protein–protein interactions that are focused upon the regulation of E2F transcription factors. In its simplest form E2F regulation consists of an interactions between E2Fs and pRB family members that suppress E2F activity and potentially repress entire E2F-regulated promoters in the G_0 and early G_1 phases of the cell cycle. With the accumulation of sufficient mitogenic signals, the cell's cell cycle control machinery in the form of cyclin/cdk kinases phosphorylate pRB and related proteins in late G_1, and thereby release free E2F and permit the expression of genes that promote cell cycle progression. This simple kind of regulation may fully characterize E2F control in *Drosophila* cells, where only single E2F, DP, and *Rb*-like genes are known to exist.

The situation is far more complicated in mammalian cells, where multiple E2Fs and pRB-like proteins interact in a variety of ways, some of which are of unknown significance. This expanded network of interactions may have evolved in order to fine tune the transcriptional control of cell growth, with different E2F multiprotein complexes coordinately regulating different groups of genes, or perhaps regulating the same genes in response to different extra-cellular signals or in distinct cell types. However, even these scenarios provide no rationale for the occurrence of the large four part E2F complexes typically present in S phase cells or the even more elaborate complexes that may form during differentiation and senescence. Clearly there is a long way to go in deciphering how E2F regulation contributes to cell cycle control.

Acknowledgments. D.C. was supported by the Susan G. Komen Breast Cancer Foundation during the preparation of this review. C.A. Afshari, R. Bernards, S. Chellappan, D. Dean, N. Dyson, D. Ginsberg, N. Hay, R. Herrera, Y. Geng, W. Krek, G. Vairo, and L. Zhu are thanked for communicating results prior to publication. N. Dyson, R.A. Weinberg, and members of the Weinberg laboratory are thanked for critical reading of the manuscript.

References

Afshari CA, Nichols MA, Xiong Y, Mudryj M (1995) The p21 protein interacts with multiple E2F complexes during the cell cycle in normal human fibroblasts (submitted)

Arroyo M, Bagchi S, Raychaudhuri P (1993) Association of the human papillomavirus type 16 E7 protein with the S-phase-specific E2F-cyclin A complex. Mol Cell Biol 13: 6537–6546

Bagchi S, Raychaudhuri P, Nevins JR (1990) Adenovirus E1A proteins can dissociate heteromeric complexes involving the E2F transcription factor: a novel mechanism for E1A trans-activation. Cell 62: 659–669

Bagchi S, Weinmann R, Raychaudhuri P (1991) The retinoblastoma protein copurifies with E2F-I, an E1A-regulated inhibitor of the transcription factor E2F. Cell 65: 1063–1072

Bandara LR, La Thangue NB (1991) Adenovirus E1A prevents the retinoblastoma gene product from complexing with a cellular transcription factor. Nature 351: 494–497

Bandara LR, Adamczewski JP, Hunt T, La Thangue NB (1991) Cyclin A and the retinoblastoma gene product complex with a common transcription factor. Nature 352: 249–251

Bandara LR, Lam EWF, Sorensen TS, Zamanian M, Girling R, La Thangue MB (1994) Dp-1: a cell cycle-regulated and phosphorylated component of transcription factor DRTF1/E2F which is functionally important for recognition by pRb and the adenovirus E4 orf 6/7. EMBO J 13: 3104–3114

Barbeau D, Charbonneau R, Whalen SG, Bayley ST, Branton PE (1994) Functional interactions within E1A protein complexes. Oncogene 9: 359–373

Beijersbergen RL, Kerkhoven RM, Zhu L, Carlee L, Voorhoeve PM, Bernards R (1994) E2F-4, a new member of the E2F gene family, has oncogenic activity and associates with p107 in vivo. Genes Dev 8: 2680–2690

Beijersbergen RL, Carlee L, Kerkhoven RM, Bernards R (1995) Regulation of the retinoblastoma protein-related p107 by G1 cyclin complexes. Genes Dev 9: 1340–1353

Bernards R, Shackleford GM, Gerber MR, Horowitz JM, Friend SH, Schartl M, Bogenmann E, Rapaport JM, McGee T, Dryja TP, Weinberg RA (1989) Structure and expression of the murine retinoblastoma gene and characterization of its encoded protein. Proc Natl Acad Sci USA 86: 6474–6478

Blake MC, Azizkhan JC (1989) Transcription factor E2F is required for efficient expression of the hamster dihydrofolate reductase gene in vitro and in vivo. Mol Cell Biol 9: 4994–5002

Buchkovich K, Duffy LA, Harlow E (1989) The retinoblastoma protein is phosphorylated during specific phases of the cell cycle. Cell 58: 1097–1105

Cao L, Faha B, Dembski M,L-HT, Harlow E, Dyson N (1992) Independent binding of the retinoblastoma protein and p107 to the transcription factor E2F. Nature 355: 176–179

Carlotti F, Crawford L (1993) Trans-activation of the adenovirus E2 promoter by human papillomavirus type 16 E7 is mediated by retinoblastoma-dependent and -independent pathways. J Gen Virol 74: 2479–2486

Chellappan S, Kraus VB, Kroger B, Munger K, Howley PM, Phelps WC, Nevins JR (1992) Adenovirus E1A, simian virus 40 tumor antigen, and human papillomavirus E7 protein share the capacity to disrupt the interaction between transcription factor E2F and the retinoblastoma gene product. Proc Natl Acad Sci USA 89: 4549–4593

Chellappan SP, Hiebert S, Mudryj M, Horowitz JM, Nevins JR (1991) The E2F transcription factor is a cellular target for the pRB protein. Cell 65: 1053–1061

Chen P-L, Scully P, Shew J-Y, Wang JYJ, Lee W-H (1989) Phosphorylation of the retinoblastoma gene product is modulated during the cell cycle and cellular differentiation. Cell 58: 1193–1198

Chittenden T, Livingston DM, Kaelin WG (1991) The T/E1A-binding domain of the retinoblastoma product can interact selectively with a sequence-specific DNA-binding protein. Cell 65: 1073–1082

Chittenden T, Livingston DM, DeCaprio JA (1993) Cell cycle analysis of E2F in primary human T cells reveals novel E2F complexes and biochemically distinct forms of free E2F. Mol Cell Biol 13: 3975–3983

Clarke AR, Maandag ER, van Roon M, van der Lugt NMT, van der Valk M, Hooper ML, Berns A, te Riele H (1992) Requirement for a functional pRB-1 gene in murine development. Nature 359: 328–330

Claudio PP, Howard CM, Baldi A, Luca AD, Fu Y, Condorelli G, Sun Y, Colburn N, Calabretta B, Giordano A (1994) p130/pRb2 has growth suppressive properties similar to yet distinctive from those of retinoblastoma family members pRb and p107. Cancer Res 54: 5556–5560

Cobrinik D, Whyte P, Peeper DS, Jacks T, Weinberg RA (1993) Cell cycle–specific association of E2F with the p130 E1A binding protein. Genes Dev 7: 2392–2404

Cress WD, Johnson DG, Nevins JR (1993) A genetic analysis of the E2F1 gene distinguishes regulation by pRB, P107, and adenovirus E4. Mol Cell Biol 13: 6314–6325

Dalton S (1992) Cell cycle regulation of the human cdc2 gene. EMBO J 11: 1797–1804

DeCaprio JA, Ludlow JW, Lynch D, Furukawa Y, Griffin J, Piwnica-Worms H, Huang C-M, Livingston DM (1989) The product of the retinoblastoma susceptibility gene has properties of a cell cycle regulatory element. Cell 58: 1085–1095

DeCaprio JA, Furukawa Y, Achenbaum F, Griffin JD, Livingston DM (1992) The retinoblastoma-susceptibility gene product becomes phosphorylated in multiple stages during cell cycle entry and progression. Proc Acad Sci USA 89: 1795–1798

Devoto SH, Mudryj M, Pines J, Hunter T, Nevins JR (1992) A cyclin A-protein kinase complex possesses sequence-specific DNA binding activity: p33cdk2 is a component of the E2F-cyclin A complex. Cell 68: 167–176

Dowdy SF, Hinds PW, Louie K, Reed SI, Arnold A, Weinberrg RA (1993) Physical interaction of the retinoblastoma protein with human D cyclins. Cell 73: 499–511

Dulic V, Lees E, Reed SI (1992) Association of human cyclin E with a periodic G1-S phase protein kinase. Science 257: 1958–1961

Durfee T, Becherer K, Chen P, Yeh S, Yang Y, Kilburn AE, Lee W, Elledge SJ (1993) The retinoblastoma protein associates with the protein phosphatase type 1 catalytic subunit. Genes Dev 7: 555–569

Dynlacht BD, Brook A, Dembski M, Yenush L, Dyson N (1994a) DNA-binding and trans-activation properties of Drosophila E2F and DP proteins. Proc Natl Acad Sci USA 91: 6359–6363

Dynlacht BD, Flores O, Lees JA, Harlow E (1994b) Differential regulation of E2F trans-activation by cyclin/cdk2 complexes. Genes Dev 8: 1772–1786

Dyson N, Guida P, Munger K, Harlow E (1992) Homologous sequences in adenovirus E1A and human papillomavirus E7 proteins mediate interaction with the same set of cellular proteins. J Virol 66: 6893–6902

Dyson N, Dembski M, Fattaey A, Ngwu C, Ewen M, Helin K (1993) Analysis of p107-associated proteins: p107 associates with a form of E2F that differs from pRB-associated E2F-1. J Virol 67: 7641–7647

Egan C, Bayley ST, Branton PE (1989) Binding of the Rb1 protein to E1A products is required for adenovirus transformation. Oncogene 4: 383–388

Ewen ME, Xing Y, Lawrence JB, Livingston DM (1991) Molecular cloning, chromosomal mapping, and expression of the cDNA for p107, a retinoblastoma gene product-related protein. Cell 66: 1155–1164

Ewen ME, Faha B, Harlow E, Livingston D (1992) Interaction of p107 with cyclin A independent of complex formation with viral oncoproteins. Science 255: 85–87

Ewen ME, Sluss HK, Sherr CJ, Matsushime H, Kato J-Y, Livingston DM (1993) Functional interactions of the retinoblastoma protein with mammalian D-type cyclins. Cell 73: 487–497

Fagan R, Flint KJ, Jones N (1994) Phosphorylation of E2F-1 modulates its interaction with the retinoblastoma gene product and the adenoviral E4 19 kDa protein. Cell 78: 799–811

Faha B, Ewen ME, Tsai L-H, Livingston DM, Harlow E (1992) Interaction between human cyclin A and adenovirus E1A-associated p107 protein. Science 255: 87–90

Faha B, Harlow E, Lees E (1993) The adenovirus E1A-associated kinase consists of cyclin e-p33cdk2 and cyclinA-p33cdk2. J Virol 67: 2456–2465

Fattaey AR, Harlow E, Helin K (1993a) Independent regions of adenovirus E1A are required for binding to and dissociation of E2F-protein complexes. Mol Cell Biol 13: 7267–7277

Fattaey AR, Helin K, Dembski MS, Dyson N, Harlow E, Vuocolo GA, Hanobik MG, Haskell KM, Oliff A, Defeo-Jones D, Jones RE (1993b) Characterization of the retinoblastoma binding proteins RBP1 and RBP2. Oncogene 8: 3149–3156

Flemington EK, Speck SH, Kaelin WG (1993) E2F-1 mediated trans-activation is inhibited by complex formation with the retinoblastoma susceptibility gene product. Proc Natl Acad Sci USA 90: 6914–6918

Fung Y, T'Ang A, Murphree AL, Zhang F, Qiu W, Wang S, Shi X, Lee L, Driscoll B, Wu K (1993) The pRB gene suppresses the growth of normal cells. Oncogene 8: 2659–2672

Ginsberg D, Vairo G, Chittenden T, Xiao Z, Xu G, Wydner KL, DeCaprio JA, Lawrence JB, Livingston DM (1994) E2F-4, a new member of the E2F transcription factor family, interacts with p107. Genes Dev 8: 2665–2679

Goodrich DW, Wang NP, Qian Y-W, Lee EY-HP, Lee WH (1991) The retinoblastoma gene product regulates progression through the G_1 phase of the cell cycle. Cell 67: 293–302

Gu W Schneider JW, Condorelli G, Kaushal S, Mahdavi V, Nadal-Ginard B (1993) Interaction of myogenic factors and the retinoblastoma protein mediates muscle cqcell commitment and differentiation. Cell 72: 309–324

Hagemeier C, Bannister AJ, Cook A, Kouzarides T (1993a) The activation domain of transcription factor PU.1 binds the retinoblastoma (RB) protein and the transcription factor TFIID in vitro: RB shows sequence similarity to TFIID and TFIIB. Proc Natl Acad Sci USA 90: 1580–1584

Hagemeier C, Cook A, Kouzarides T (1993b) The retinoblastoma protein binds E2F residues required for activation in vivo and TBP binding in vitro. Nucleic Acids Res 21: 4998–5004

Hamel PA, Gill RM, Phillips RA, Gallie BL (1992). Transcriptional repression of the E2-containing promoters EllaE, c-myc and RBI by the product of the RB1 gene. Mol Cell Biol 12: 3431–3438

Hannon GJ, Demetrick D, Beach D (1993) Isolation of the pRB-related p130 through its interaction with CDK2 and cyclins. Genes Dev 7: 2378–2391

Harlow E, Whyte P, Franza BJ, Schley C (1986) Association of adenovirus early-region 1A proteins with cellular polypeptides. Mol Cell Biol 6: 1579–1589

Hatakeyama M, Brill JA, Fink GR, Weinberg RA (1994) Collaboration of G_1 cyclins in the functional inactivation of the retinoblastoma protein. Genes Dev 8: 1759–1771

Helin K, Harlow E (1994) Heterodimerization of the transcription factor E2F-1 and DP-1 is required for binding to the adenovirus E4(ORF6/7) protein. J Virol 68: 5027–5035

Helin K, Lees JA, Vidal M, Dyson N, Harlow E, Fattaey A (1992) A cDNA encoding a pRB-binding protein with properties of the transcription factor E2F. Cell 70: 337–350

Helin K, Harlow E, Fattaey AR (1993a) Inhibition of E2F-1 trans-activation by direct binding of the retinoblastoma protein. Mol Cell Biol 13: 6501–6508

Helin K, Wu C-L, Fattaey AR, Lees JA, Dynlacht BD, Ngwu C, Harlow E (1993b) Heterodimerization of the transcription factors E2F-1 and DP-1 leads to cooperative trans-activation. Genes Dev 7: 1850–1861

Herrera RE, Tan VP, Williams BO, Weinberg RA, Jacks T (1995) Altered cell cycle kinetics and gene expression in Rb-deficient fibroblasts (submitted)

Herrmann C, Su L, Harlow E (1991) Adenovirus E1A is associated with a serine/threonine protein kinase. J Virol 65: 5848–5859

Hiebert SW (1993) Regions of the retinoblastoma gene product required for its interaction with the E2F transcription factor are necessary for E2 promoter repression and pRb-mediated growth suppression. Mol Cell Biol 13: 3384–3391

Hiebert SW, Lipp M, Nevins JR (1989) E1A-dependent trans-activation of the human myc promoter is mediated by the E2F factor. Proc Natl Acad Sci USA 86: 3594–3598

Hiebert SW, Chellappan SP, Horowitz JM, Nevins JR (1992) The interaction of pRB with E2F coincides with an inhibition of the transcriptional activity of E2F. Genes Dev 6: 177–185

Hinds PW, Mittnacht S, Dulic V, Arnold A, Reed SI, Weinberg RA (1992) Regulation of retinoblastoma protein functions by ectopic expression of human cyclins. Cell 70: 993–1006

Hsiao K-M, McMahon SL, Farnham PJ (1994) Multiple DNA elements are required for the growth regulation of the mouse E2F1 promoter. Genes Dev 8: 1526–1537

Huang S, Wang N, Tseng BY, Lee W-H, Lee EY-HP (1990) Two distinct and frequently mutated regions of retinoblastoma protein are required for binding to SV40 T antigen. EMBO J 9: 1815–1822

Huang S, Shin E, Sheppard K, Chokroverty L, Shan B, Qian Y, Lee EY, Lee W (1992) The retinoblastoma region required for interaction with the E2F transcription factor includes the T/E1A binding and carboxy-terminal sequences. DNA Cell Biol 11: 539–548

Hunter T, Pines J (1991) Cyclins and cancer. Cell 66: 1071–1074

Ikeda M, Nevins JR (1993) Identification of distinct roles for separate E1A domains in disruption of E2F complexes. Mol Cell Biol 13: 7029–7035

Imai Y, Matsushima Y, Sugimura T, Terada M (1991) Purification and characterization of human papilloma virus type 16 E7 protein with preferential binding capacity to the underphosphorylated form of retinoblastoma gene product. J Virol 65: 4966–4972

Ivey-Hoyle M, Conroy R, Huber HE, Goodhart PJ, Oliff A, Heimbrook DC (1993) Cloning and characterization of E2F-2, a novel protein with the biochemical properties of transcription factor E2F. Mol Cell Biol 13: 7802–7812

Jacks T, Fazeli A, Schmitt EM, Bronson RT, Goodell MA, Weinberg RA (1992) Effects of an pRB mutation in the mouse. Nature 359: 295–300

Jiang H, Lin J, Young S-M, Goldstein NI, Waxman S, Davilla V, Chellappan SP, Fisher PB (1995) Cell cycle gene expression and E2F transcription factor complexes in human melanoma cells introduced to terminally differentiate. Oncogene (in press)

Johnson GD, Ohtani K, Nevins JR (1994) Autoregulatory control of E2F1 expression in response to positive and negative regulators of cell cycle progression. Genes Dev 8: 1514–1525

Kaelin WG, Pallas DC, DeCaprio JA, Kaye FJ, Livingston DM (1991) Identification of cellular proteins that can interact specifically with the T/E1A-binding region of the retinoblastoma gene product. Cell 64: 521–532

Kaelin WG, Krek W, Sellers WR, DeCaprio JA, Ajchenbaum F, Fuchs CS, Chittenden T, Li Y, Farnham PJ, Blanar MA, Livingston DM, Flemington EK (1992) Expression cloning of a cDNA encoding a retinoblastoma-binding protein with E2F-like properties. Cell 70: 351–364

Kato J, Matsushime H, Hiebert SW, Ewen ME, Sherr CJ (1993) Direct binding of cyclin D to the retinoblastoma gene product (pRb) and pRb phosphorylation by the cyclin D-dependent Kinase, CDK4. Genes Dev 7: 331–342

Kiyokawa H, Richon VM, Rifkind RA, Marks PA (1994) Suppression of cyclin-dependent kinase 4 during induced differentiation of erythroleukemia cells. Mol Cell Biol 14: 7195–7203

Kovesdi I, Reichel R, Nevins JR (1986) Identification of a cellular transcription factor involved in E1A trans-activation. Cell 45: 219–228

Kovesdi I, Reichel R, Nevins JR (1987) Role of an adenovirus E2 promoter binding factor in E1A-mediated coordinate gene control. Proc Natl Acad Sci USA 84: 2180–2184

Krek W, Livingston DM, Shirodkar S (1993) Binding to DNA and the retinoblastoma gene product promoted by complex formation of different E2F family members. Science 262: 1557–1560

Krek W, Ewen ME, Shirodkar S, Arany Z, Kaelin WG, Livingston DM (1994) Negative Regulation of the growth-promoting transcription factor E2F-1 by a stably bound cyclin A-dependent protein kinase. Cell 78: 161–172

Lam EW, Watson RJ (1993) An E2F-binding site mediates cell-cycle regulated repression of mouse B-myb transcription. EMBO J 12: 2705–2713

Lam EW, Morris JDH, Davies R, Crook T, Watson RJ, Vousden KG (1994) HPV E7 oncoprotein deregulates B-myb expression: correlation with targeting of p107/E2F complexes. EMBO J 13: 871–878

La Thangue NB, Rigby PWJ (1987) An adenovirus E1A-like transcription factor is regulated during the differentiation of murine embryonal carcinoma cells. Cell 49: 507–513

La Thangue NB, Thimmappaya B, Rigby PWJ (1990) The embryonal carcinoma stem cell E1a-like activity involves a differentiation-regulated transcription factor. Nucleic Acids Res 18: 2929–2938

Lees E, Faha B, Dulic V, Reed SI, Harlow E (1992) Cyclin E/cdk2 and cyclin A/cdks kinases associate with p107 and E2F in a temporally distinct manner. Genes Dev 6: 1874–1885

Lee EY, Chang C, Hu N, Wang Y, Lai C, Herrup K, Lee W, Bradley A (1992) Mice deficient for pRB are nonviable and show defects in neurogenesis and haematopoiesis. Nature 359: 288–294

Lees J, Saito M, Vidal M, Valentino M, Look T, Harlow E, Dyson N, Helin K (1993) The retinoblastoma protein binds to a family of E2F transcription factors. Mol Cell Biol 13: 7813–7825

Lees LA, Buchkovich KJ, Marshak DR, Anderson CW, Harlow E (1991) The retinoblastoma protein is phosphorylated on multiple sites by human cdc2. EMBO J 10: 4279–4290

Lew DJ, Dulic V, Reed SI (1991) Isolation of three novel human cyclins by rescue of G_1 cyclin (Cln) function in yeast. Cell 66: 1197–1206

Li Y, Graham C, Lacy S, Duncan AMV, Whyte P (1993) The adenovirus E1A-associated 130 kD protein is encoded by a member of the retinoblastoma gene family and physically interacts with cyclins A and E. Genes Dev 7: 2366–2377

Li Y, Slansky JE, Myers DJ, Drinkwater NR, Kaelin WG, Farnham PJ (1994) Cloning, Chromosomal location, and characterization of mouse E2F-1. Mol Cell Biol 14: 1861–1869

Lin BT-Y, Gruenwald S, Morla AO, Lee W-H, Wang JYJ (1991) Retinoblastoma cancer suppressor gene product is a substrate of the cell cycle regulator cdc2 kinase. EMBO J 10: 857–864

Ludlow JW, DeCaprio JA, Huang C-M, Lee W-H, Paucha E, Livingston DM (1989) SV40 large T antigen binds preferentially to an underphosphorylated member of the retinoblastoma susceptibility gene product family. Cell 56: 57–65

Ludlow JW, Shon J, Pipas JM, Livingston DM, DeCaprio JA (1990) The retinoblastoma susceptibility gene product undergoes cell cycle-dependent dephosphorylation and binding to and release from SV40 large T. Cell 60: 387–396

Melamed D, Tiefenbrun N, Yarden A, Kimchi A (1993) Interferons and Interleukin-6 suppress the DNA-binding activity of E2F in growth-sensitive hematopoietic cells. Mol Cell Biol 13: 5255–5265

Mihara K, Cao X-R, Yen A, Chandler S, Driscoll B, Murphree AL, Tàng A, Fung Y-KT (1989) Cell cycle-dependent regulation of phosphorylation of the human retinoblastoma gene product. Science 246: 1300–1303

Mittnacht S, Lees JA, Desai D, Harlow E, Morgan DO, Weinberg RA (1994) Distinct sub-populations of the retinoblastoma protein show a distinct pattern of phosphorylation. EMBO J 13: 118–127

Mudryj M, Devoto SH, Hiebert SW, Hunter T, Pines J, Nevins JR (1991) Cell cycle regulation of the E2F transcription factor involves an interaction with cyclin A. Cell 65: 1243–1253

Neuman E, Flemington EK, Sellers WR, Kaelin WG (1994) Transcription of the E2F1 gene is rendered cell cycle-dependent by E2F DNA bindidng sites within its promoter. Mol Cell Biol 14: 6607–6615

Ohtani K, Nevins JR (1994) Functional properties of a Drosophila homologue of the E2F1 gene. Mol Cell Biol 14: 1603–1612

Pagano M, Draetta G, and Jansen-Durr P (1992a) Association of cdk2 kinase with the transcription factor E2F during S phase. Science 255: 1144–1147

Pagano M, Durst M, Joswig S, Draetta G, Jansen-Durr P (1992b) Binding of the human E2F transcription factor to the retinoblastoma protein but not to cyclin A is abolished in HPV-16-immortalized cells. Oncogene 7: 1681–1686

Partridge JF, La Thangue NB (1991) A developmentally regulated and tissue-dependent transcription factor complexes with the retinoblastoma gene product. EMBO J 10: 3819–3827

Peeper DS, Parker LL, Ewen ME, Toebes M, Hall FL, Xu M, Zantema A, van den Eb AJ, Piwnica-Worms H (1993) A- and B-type cyclins differentially modulate substrate specificity of cyclin-cdk complexes. EMBO J 12: 1947–1954

Philpott A, Friend SH (1994) E2F and its developmental regulation in *Xenopus* laevis. Mol Cell Biol 14: 5000–5009

Qian Y, Luckey C, Horton L, Esser M, Templeton DJ (1992) Biological function of the retinoblastoma protein requires distinct domains for hyperphosphorylation and transcription factor binding. Mol Cell Biol 12: 5363–5372

Qian Y, Wang Y, Hollingsworth RE, Jones D, Ling N, Lee EY (1993) A retinoblastoma-binding protein related to a negative regulator of ras in yeast. Nature 364: 648–652

Qin X, Chittenden T, Livingston DM, Kaelin WG (1992) Identification of a growth suppression domain within the retinoblastoma gene product. Genes Dev 6: 953–964

Qin X, Livingston DM, Ewen M, Sellers WR, Arany Z, Kaelin WG (1995) The transcription factor E2F-1 is a downstream target of pRB action. Mol Cell Biol 15: 742–755

Ray SK, Arroyo M, Bagchi S, Raychaudhuri P (1992) Identification of a 60-kilodalton Rb-binding protein, RBP60, that allow the Rb-E2F complex to bind DNA. Mol Cell Biol 12: 4327–4333

Raychaudhuri P, Bagchi S, Devoto SH, Kraus VB, Moran E, Nevins JR (1991) Domains of the adenovirus E1A protein required for oncogenic activity are also required for dissociation of E2F transcription factor complexes. Genes Dev 5: 1200–1211

Reed SI (1992) The role of p34 kinases in the G_1 to S phase transition. Annu Rev Cell Biol 8: 529–561

Reichel R, Kovesdi I, Nevins JR (1987) Developmental control of a promoter-specific factor that is also regulated by the E1A gene product. Cell 48: 501–506

Reichel RR (1992) Regulation of E2F/cyclin A-containing complex upon retinoic acid-induced differentiation of teratocarcinoma cells. Gene Exp 2: 259–271

Robbins PD, Horowitz JM, Mulligan RC (1990) Negative regulation of human c-fos expression by the retinoblastoma gene product. Nature 346: 668–671

Rustgi AK, Dyson N, Bernards R (1991) Amino-terminal domains of c-myc and N-myc proteins mediate binding to the retinoblastoma gene product. Nature 352: 541–544

Sardet C, Vidal M, Cobrinik D, Geng Y, Onufryk C, Chen A, Weinberg RA (1995) E2F-4 and E2F-5, two novel members of the E2F family, are expressed in the early phases of the cell cycle. Proc Natl Acad Sci USA 92: 2403–2407

Savatier P, Huang S, Szekely L, Wiman KG, Samarut J (1994) Contrasting patterns of retinoblastoma protein expression in mouse embryonic stem cells and embryonic fibroblasts. Oncogene 9: 809–818

Schwarz JK, Devoto SH, Smith EJ, Chellappan SP, Jakoi L, Nevins JR (1993) Interactions of the p107 and pRB proteins with E2F during the cell proliferation response. EMBO J 12: 1013–1020

Shan B, Zhu X, Chen P-L, Durfee T, Yang Y, Sharp D, Lee W-H (1992) Molecular cloning of cellular genes encoding retinoblastoma-associated proteins: identification of a gene with properties of the transcription factor E2F. Mol Cell Biol 12: 5620–5631

Shirodkar S, Ewen M, DeCaprio JA, Morgan J, Morgan DM, Chittenden T (1992) The transcription factor E2F interacts with the retinoblastoma product and a p107-cyclin A complex in a cell cycle-regulated manner. Cell 68: 157–166

Slansky JE, Li Y, Kaelin WG, Farnham PJ (1993) A protein synthesis-dependent increase in E2F1 mRNA correlates with growth regulation of the dihydrofolate reductase promoter. Mol Cell Biol 13: 1610–1618

Smith EJ, Nevins JR (1995) The pRB-related p107 protein can suppress E2F function independently of binding to cyclin A/cdk2. Mol Cell Biol 15: 338–344

Szekely L, Jiang W-Q, Bulic-Jakus F, Rosen A, Ringertz N, Klein G, Wiman KG (1992) Cell type and differentiation dependent heterogeneity in retinoblastoma protein expression in SCID mouse fetuses. Cell Growth Differ 3: 149–156

Templeton D, Park SH, Lanier L, Weinberg RA (1991) Nonfunctional mutants of the retinoblastoma protein are characterized by defects in phosphorylation, viral oncoprotein association, and nuclear tethering. Proc Natl Acad Sci USA 88: 3033–3037

Thalmeier K, Synovzik H, Mertz R, Winnacker E-L, Lipp M (1989) Nuclear factor E2F mediates basic transcription and trans-activation by E1A of the human myc promoter. Genes Dev 3: 527–536

Thut CJ, Chen J, Klemm R, Tjian R (1995) p53 transcriptional activation mediated by coactivators TAFII40 and TAFII60. Science 267: 100–104

Vairo G, Livingston DM, Ginsberg D (1995) Functional interaction between E2F-4 and p130: evidence for distinct mechanisms underlying growth suppression by different retinoblastoma protein family members. Genes Dev 9: 869–881

Wang C, Petryniak B, Thompson CB, Kaelin WG, Leiden JM (1993) Regulation of the Ets-related transcription factor Elf-1 by binding to the retinoblastoma protein. Science 260: 1330–1335

Weinberg RA (1995) The Rb protein and cell cycle control. Cell 81: 323–330

Weintraub SJ, Prater CA, Dean DC (1992) Retinoblastoma protein switches the E2F site from positive to negative element. Nature 358: 259–262

Weintraub SJ, Chow KNB, Luo RX, Zhang SH, He S, Dean DC (1995) Mechanism of active transcriptional repression by the retinoblastoma protein . Nature 375: 812–815

Welch PJ, Wang JYJ (1993) A C-terminal protein binding domain in pRB regulates the nuclear c-Abl tyrosine kinase in the cell cycle. Cell 75: 779–790

Welch PJ, Wang JYJ (1995) Disruption of retinoblastoma protein function by coexpression of its C pocket fragment. Genes Dev 9: 31–46

Whyte P, Buchkovich KJ, Horowitz JM, Friend SH, Raybuck M, Weinberg RA, Harlow E (1988) Association between an oncogene and an anti-oncogene: the adenovirus E1A proteins bind to the retinoblastoma gene product. Nature 334: 124–129

Whyte P, Williamson NM, Harlow E (1989) Cellular targets of transformation by the adenovirus E1A protein. Cell 56: 67–75

Wu EW, Clemens KE, Heck DV, Munger K (1993) The human papillomavirus E7 oncoprotein and the cellular transcription factor E2F bind to separate sites on the retinoblastoma tumor suppressor protein. J Virol 67: 2402–2407

Xu M, Sheppard K, Peng C, Yee AS, Piwnica-Worms H (1994) Cyclin A/cdk2 binds directly to E2F-1 and inhibits the DNA-binding activity of E2F-1/DP-1 by phosphorylation. Mol Cell Biol 14: 8420–8431

Yee S.-Y, Branton PE (1985) Detection of cellular proteins associated with human adenovirus type 5 early region 1A polypeptides. Virol 147: 142–153

Zamanian M, La Thangue NB (1992) Adenovirus E1a prevents the retinoblastoma gene product from repressing the activity of a cellular transcription factor. EMBO J 11: 2603–2610

Zamanian M, La Thangue NB (1993) Transcriptional repression by the pRB-related protein p107. Mol Biol Cell 4: 389–396

Zhu L, van den Heuvel S, Helin K, Fattaey A, Ewen M, Livingston D, Dyson N, Harlow E (1993) Inhibition of cell proliferation by p107, a relative of the retinoblastoma protein. Genes Dev 7: 1111–1125

Zhu L, Enders G, Lees AJ, Beijersbergen RL, Bernards R, Harlow E (1995) The pRB-related p107 contains two growth suppression domains: independent interactions with E2F and cyclin/cdk complexes. EMBO J (in press)

Use of the E2F Transcription Factor by DNA Tumor Virus Regulatory Proteins

W.D. Cress and J.R. Nevins

1 Introduction

In most cases the normal host cell for infection by the DNA tumor virus is a quiescent, terminally differentiated cell that is not dividing. Various experiments have demonstrated that upon infection these cells are stimulated to enter S phase, as indicated by the synthesis of cellular DNA and the induction of activities associated with DNA replication, particularly those enzymes involved in deoxynucleotide biosynthesis (Hatanaka and Dulbecco 1966; Ledinko 1968; Yamashita and Shimojo 1969; Dulbecco et al. 1965; Frearson et al. 1965, 1966; Hartwell et al. 1965; Kara and Weil 1967; Kit et al. 1966a, b, 1967a, b; Sheinin 1966). This viral-mediated S phase induction almost certainly reflects the need of these viruses to create an *environment* appropriate for viral DNA synthesis since the levels of deoxynucleotides are low in quiescent cells and normally rise only when cells are stimulated to enter S phase (Bjorklund et al. 1990; Engstrom et al. 1985). Thus, the normal host for infection by these viruses, a nongrowing cell, is not an environment conducive to DNA replication. The capacity of the DNA tumor viruses to drive a quiescent cell into S phase is dependent largely on the action of the viral regulatory proteins that include adenovirus E1A, SV40 T antigen, and human

Department of Genetics, Howard Hughes Medical Institute, Duke University Medical Center, Durham, NC 27710, USA papillomavirus (HPV) E7. These are also viral proteins that possess oncogenic activity through their common ability to inactivate the retinoblastoma gene product Rb. Indeed, it is now clear that the ability of these viral proteins to promote entry into S phase, so as to create an environment that facilitates viral DNA replication, also results in a loss of cell growth control when a viral infection cannot proceed to completion. Recent developments have led to the realization that these viral proteins mediate these events through the activation of the E2F transcription factor, and studies of their interactions have provided considerable insight into the basic mechanisms of cell growth control and oncogenesis.

2 Common Targets for the DNA Tumor Virus Oncoproteins

Much of what is known about the activities of these DNA tumor virus proteins has been inferred from the physical associations of these viral proteins with a variety of cellular polypeptides, focused initially on E1A interactions (YEE and BRANTON 1985; HARLOW et al. 1986). Coimmunoprecipitation experiments identified at least six cellular proteins, ranging in molecular weight from 300 to 33 kDa, that were in stable complexes with E1A (YEE and BRANTON 1985; HARLOW et al. 1986). Subsequent work identified one of these proteins as the retinoblastoma (Rb) gene product (WHYTE et al. 1988), a finding of profound importance with respect to the oncogenic action of E1A. Further work has shown that two additional E1A-associated proteins are related to Rb (p130 and p107; ZHU et al. 1993; EWEN et al. 1991; LI et al. 1993; HANNON et al. 1993; MAYOL et al. 1993), two are cell cycle regulatory proteins (cyclin A and cdk2; TSAI et al. 1991), and that the largest E1A associated protein, p300, may function as a transcription factor (ECKNER et al. 1994). It soon became clear that each of the DNA tumor viruses encode a protein that can bind to the Rb tumor suppressor protein (Table 1; DYSON et al. 1989; DECAPRIO et al. 1988). Thus, although the adenoviruses, polyomaviruses, and papillomaviruses share few structural relationships, the regulatory proteins of the three groups of viruses appear to share a common function. In addition, a sequence comparison of these viral proteins revealed a short region of homology that included the domain important for binding to Rb (FIGGE et al. 1988; Fig. 1).

In addition to the ability of DNA tumor virus proteins to bind to and inactivate Rb, each of these viruses also encodes a protein that interacts with the p53 tumor suppressor and, as a consequence of this interaction, inactivates p53 function (Table 1). In contrast to the targeting of Rb, however, there is no apparent conservation of viral protein sequences involved in these interactions. Moreover, it appears that distinct domains of the p53 protein are recognized by the viral proteins, and the immediate consequence of this interaction is different. For instance, whereas the E1B 55-kDa protein binds to p53 and apparently blocks its

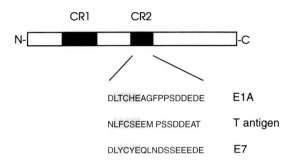

Fig. 1. Homology in the viral sequences involved in binding to the Rb family of proteins. The regions in E1A that exhibit homology with sequences in SV40 T antigen and HPV E7 are depicted (FIGGE et al. 1988). *Shaded boxes*, the L-X-C-X-E motif found in the CR2 region of E1A, which is shared with various cellular proteins (DEFEO-JONES et al. 1991) including the D-type cyclins (MATSUSHIME et al. 1991; XIONG et al. 1991)

Table 1. Cellular targets of the DNA tumor virus oncoproteins

Virus	Cellular target
Adenovirus	
E1A	Rb
E1B$_{19K}$	?
E1B$_{55K}$	p53
SV40	
Large T antigen	Rb, p53
Polyoma	
Large T antigen	Rb
Middle T antigen	c-src
Papillomavirus	
E7	Rb
E6	p53

transcriptional activating function, the interaction of the HPV E6 protein leads to the degradation of p53. Nevertheless, despite these differences the net result is the same – a loss of p53 function.

3 A Role for Rb Inactivation During Productive Infection by the DNA Tumor Viruses: Activation of the E2F Transcription Factor

Although the inactivation of Rb by E1A, T antigen, and E7 is most often viewed in the context of oncogenesis, this event must play a role in lytic growth since these viruses clearly have not evolved to be oncogenic. Rather, the true role of these viral proteins must be to facilitate the replication of the virus in its quiescent host. A connection between the action of these proteins to facilitate lytic growth, to stimulate S phase, and to act as oncogenic agents was revealed with the identification of the E2F transcription factor as a target for the Rb protein and the action of E1A. The E2F transcription factor was identified as an important component for transcription of the adenovirus E2 gene (KOVESDI et al. 1986b). Subsequent studies revealed that this cellular transcription factor is normally complexed to other cellular proteins in most cell types, that these interactions prevent the activation of E2 transcription, and that the E1A protein possesses the capacity to disrupt these complexes, releasing E2F that can be utilized for E2 transcription (BAGCHI et al. 1990). The ability of E1A to mediate this dissociation was shown to depend on viral sequences that are known to be important for binding to the Rb protein (RAYCHAUDHURI et al. 1991). This suggested a relationship between Rb binding and E2F complex dissociation, as depicted in Fig. 2, whereby if the Rb protein were a component of the E2F complex, the binding of E1A to Rb could be viewed as the result of the disruption of the E2F-Rb complex. A variety of experiments have now shown that the Rb protein is indeed a component of the E2F complexes along with the majority of the other proteins previously identified as E1A-binding proteins (BAGCHI et al. 1991; BANDARA et al. 1991; BANDARA and LA THANGUE 1991; CHELLAPPAN et al. 1991; CAO et al. 1992; DEVOTO et al. 1992; SHIRODKAR et al. 1992).

Other studies demonstrated that Rb-containing cellular protein complexes can bind to DNA, and the elucidation of the sequence specificity of these interactions revealed the E2F recognition sequence (CHITTENDEN et al. 1991). Finally, subsequent experiments demonstrated that the other DNA tumor virus proteins known to bind to the Rb protein, SV40 T antigen, and HPV E7 can also

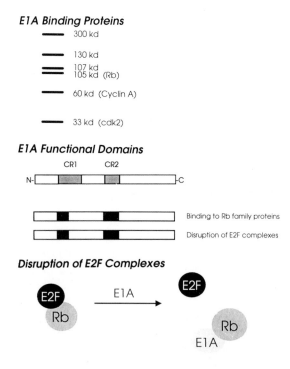

E1A Binding Proteins
—— 300 kd
—— 130 kd
≡ 107 kd
  105 kd (Rb)
—— 60 kd (Cyclin A)
—— 33 kd (cdk2)

E1A Functional Domains
CR1   CR2

Binding to Rb family proteins
Disruption of E2F complexes

Disruption of E2F Complexes
E2F Rb   E1A →   E2F   Rb E1A

Fig. 2. The relationship of E1A binding to cellular proteins and the disruption of E2F complexes. Schematic representation of the cellular proteins that are recovered in co-immunoprecipitation assays with the adenovirus E1A protein (HARLOW et al. 1986; YEE and BRANTON 1985). *Below*, schematic representation of the E1A gene indicating the positions of the CR1 and CR2 domains and the regions involved in binding to the various cellular proteins. The Rb family includes the p105 Rb protein and the p130 and p107 proteins. CR1 sequences appear to be involved in binding to both the p300 protein and the Rb family (WANG et al. 1993)

disrupt the E2F-Rb complex (CHELLAPPAN et al. 1992), indicating that the common ability of the DNA tumor virus oncoproteins E1A, T antigen, and E7 to bind to the Rb protein is a reflection of their ability to disrupt the E2F-Rb complex.

The functional significance of this action was revealed by the finding that binding of Rb to E2F inhibits the transcriptional activation capacity of the E2F factor (HIEBERT et al. 1992; HIEBERT 1993; ZEMANIAN and LA THANGUE 1993; HAGEMEIER et al. 1993; FLEMINGTON et al.1993; HELIN et al. 1993a; CRESS et al. 1993). Thus, the action of the viral oncoproteins in releasing E2F from the inhibitory complex is one that results in the activation of E2F transcription function. This activation has direct benefit for adenovirus transcription since E2F is utilized for transcription of the viral E2 gene (KOVESDI et al. 1986a, b). In contrast, there are no E2F binding sites in either the SV40 genome or the papillomavirus genome and thus the activation of E2F by T antigen and the E7 protein cannot directly benefit these viruses. The major consequence of this activation appears to be the activation of cellular genes that encode the activities important for DNA synthesis in S phase.

The study of E2F function has provided a connection between the action of the viral oncoproteins and the ability of these viruses to drive quiescent cells into S phase. A number of experiments using the recently cloned E2F family members (KAELIN et al. 1992; GIRLING et al. 1993; SHAN et al. 1992; HELIN et al. 1992, 1993b) have now pointed to the critical importance of E2F as a target for the Rb protein. It was recognized, even before the E2F-Rb connection had been established, that E2F sites are found in the promoters of a number of genes which are required for S phase entry (HIEBERT et al. 1989, 1991; THALMEIER et al. 1989), leading to the suggestion that E2F may be a critical factor regulating the expression of S phase genes. This possibility was confirmed by experiments that demonstrated a clear role for E2F in activating the DHFR gene (SLANSKY et al. 1993), the B-myb gene (LAM and WATSON 1993), and the cdc2 gene (DALTON 1992). Moreover, very recent experiments have shown that the expression of the E2F1 gene product encoded by a recombinant adenovirus can induce the expression of many of these suspected targets (DEGREGORI et al. 1995). These experiments take on added significance since they measure the activation of the endogenous, chromo-somally located genes rather than transfected plasmids.

The fact that E2F can activate these genes and that E2F activity is controlled by Rb suggests an important role for E2F in G_1/S progression. Indeed, several experiments have now shown that the E2F1 cDNA, when introduced into otherwise quiescent cells, drives these cells to synthesize DNA (JOHNSON et al. 1993; WU and BERK 1988; QIN et al. 1994; WU and LEVINE 1994; SHAN and LEE 1994; KOWALIK et al. 1995). Subsequent experiments have shown that E2F1 can overcome a G_1 arrest induced by transforming growth factor-β (SCHWARZ et al. 1995), γ-irradiation (DEGREGORI et al. 1995), or various G_1 cyclin kinase inhibitors (J. DeGregori, G. Leone, and J. Nevins, in preparation). Finally, it has been demonstrated that enforced expression of E2F activity can transform immortalized REF cells (SINGH et al. 1994; GINSBERG et al. 1994) and can collabo-rate with an activated K-ras gene to transform primary rat embryo cells

(JOHNSON et al. 1994). Thus, a variety of experimental approaches demonstrate that the E2F1 transcription factor is a critical element in S phase entry, and that this is a critical target for tumor suppressor protein Rb.

4 A Mechanism for the E1A-Mediated Disruption of E2F-Rb Complexes

Considerable light has now been shed on the mechanism by which the adeno-virus E1A protein mediates a dissociation of E2F complexes to generate free *trans*-activating E2F (BAGCHI et al. 1990; BANDARA and LA THANGUE 1991). The ability of the E1A protein to dissociate E2F complexes was found to depend upon two domains which are highly conserved among adenovirus serotypes (HUANG et al. 1993; RAYCHAUDHURI et al. 1991). These two regions, termed CR1 and CR2, are also the regions of E1A known to be critical in binding to the Rb protein as well as Rb family members (WHYTE et al. 1989; see Fig. 1). Analysis of the interaction of E1A with Rb suggested that the E1A domains make separate contacts with Rb (DYSON et al. 1992). The mechanism of the E2F-Rb dissociation appears to involve these two conserved domains of the E1A protein acting in a two-step process as depicted in Fig. 3 (IKEDA and NEVINS 1993; FATTAEY et al. 1993). Perhaps the simplest view is to consider the E2F-Rb complex to be in an equilibrium state with the dissociated components, as depicted in the top diagram. A series of experiments have shown that the CR2 region binds to sequences in the Rb protein distinct from that involved in E2F binding (IKEDA and NEVINS 1993; FATTAEY et al. 1993). The CR2 domain of E1A contains a short amino acid consensus sequence Leu-X-Cys-X-Glu which is also present in SV40 T antigen and HPV E7 (FIGGE et al. 1988; Fig. 1), and specific mutagenesis has shown that this domain is important for Rb binding as well as activation of the E2F transcription factor (CORBEIL and BRANTON 1994). This sequence motif is also found in a number of cellular proteins which bind directly to Rb, including D-type cyclins (DEFEO-JONES et al. 1991; MATSUSHIME et al. 1991; XIONG et al. 1991). Thus, the CR2 domain would serve to bring the E1A protein to the E2F-Rb complex. Other experiments have shown that the CR1 domain must recognize sequences in Rb that coincide with or overlap the sequences recognized by E2F since a CR1 peptide can block E2F-Rb complex formation (RAYCHAUDHURI et al. 1991). Thus, once the CR2 domain allows the E1A protein to bind to the E2F-Rb complex, it might simply wait until the E2F-Rb complex dissociates under normal equilibrium conditions. Then, upon dissociation of the complex, the CR1 region of E1A could bind to the Rb domain involved in E2F binding thereby preventing the reassociation of E2F with Rb and thus driving the equilibrium towards the dissociated state.

Although equivalent experiments have not been performed for complexes involving E2F and Rb family members, nor for the ability of T antigen or E7 to act in this manner, it seems likely that this may be a general mechanism for the

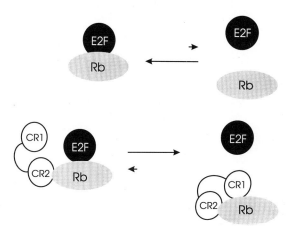

Fig. 3. The role of E1A domains in the disruption of the E2F-Rb complex. *Above*, schematic depiction of the equilibrium between the E2F-Rb complex and free E2F and free Rb; *arrows*, an equilibrium state in favor of the complex. *Below*, E1A drives the equilibrium towards the dissociated components by first interacting with the complex and then upon natural dissociation, blocks the subsequent interaction of E2F with Rb

disruption of the E2F complexes. Interestingly, this mechanism may also go beyond the action of the viral proteins. The conserved L-X-C-X-E motif, which appears to mediate the CR2-dependent interaction of E1A with the E2F-Rb complex, is also found in the N-terminal sequence of the D-type cyclin family and may mediate the binding of these cyclins to Rb (DOWDY et al. 1993). Given the fact that phosphorylation of Rb blocks its interaction with E2F (CHELLAPPAN et al. 1991), together with the fact that Rb phosphorylation is likely mediated by G_1 cyclin/kinases (KATO et al. 1993), a model as depicted in Fig. 4 becomes apparent. The D type cyclin would perform a role similar to the E1A CR2 domain, in the case bringing a kinase to the complex rather than the CR1 domain. Upon dissociation, the kinase would now be able to phosphorylate critical residues of Rb thereby blocking the reassociation of the complex.

5 A Unique Adenovirus Gene Activity that Facilitates Use of E2F for Viral Transcription

Although each of the DNA tumor viruses target E2F complexes, only the adenovirus family actually utilizes E2F for the transcription of any of its own genes. During adenovirus infection there is a temporal cascade of early gene expression involving the initial expression of the E1A and E4 transcription units very early in infection followed quickly by the E3, E1B and finally the E2 transcription unit (NEVINS 1987). The protein products of the E2 region, which include a 72-kDa single-stranded DNA binding protein, the viral DNA polymerase, and the terminal protein which primes viral DNA replication, are all expressed from a promoter located at map position 75. The E2 promoter, schematically shown in Fig. 5, contains a single ATF site, a single TATA element and two inverted repeats of the E2F recognition sequence TTTCGCGC (LOEKEN and BRADY

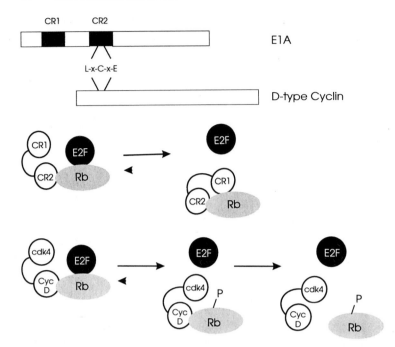

Fig. 4. A model of disruption of the E2F-Rb complex by G_1 cyclin/kinases. *Above*, the schematic highlights the homology between the E1A CR2 sequence (L-X-C-X-E) and the N terminal region of the D-type cyclins (DOWDY et al. 1993). *Below*, a schematic depiction of a potential role of the cyclin D protein and the associated cdk4 kinase in altering the equilibrium of the E2F-Rb complex

1989). These E2F sites are spaced two helical turns apart after the second C, which is likely the actual center of E2F binding as measured by DNase footprinting (YEE et al. 1989).

The activation of the E2 promoter involves not only E2F but also a 19-kDa protein product of the viral E4 6/7 ORF. The binding of a single E2F heterodimer on a single E2F site is relatively unstable, exhibiting a half-life of only a few minutes whereas the complex of E2F and E4 on the two site E2 promoter is very stable, with a half-life measured in hours (RAYCHAUDHURI et al. 1990; REICHEL et al. 1989; NEILL and NEVINS 1991; NEILL et al. 1990; OBERT et al. 1994; O'CONNOR and HEARING 1991; HUANG and HEARING 1989; HARDY and SHENK 1989; HARDY et al. 1989; MARTON et al. 1990). The formation of this stable complex requires the interaction of E4 but also requires the precise arrangement of the E2F binding sites found in the E2 promoter. Any alteration of the spacing or orientation of the binding sites virtually eliminates the formation of the E4-dependent stable complex (HARDY and SHENK 1989; RAYCHAUDHURI et al. 1990). Since the E2 promoter is the only known promoter containing this arrangement of E2F sites, these results suggest that the action of E4 converts E2F into an adenovirus E2 gene-specific transcription factor.

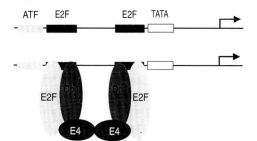

Fig. 5. Interaction of the adenovirus E4 protein with E2F. *Above*, a sehematic representation of the E2 promoter including the TATA sequence, two E2F recognition sequences, and the ATF recognition sequence. *Below*, interaction of two E2F/DP1 heterodimers with the E2F recognition sequences stabilized by the E4 homodimer

Recent experiments utilizing cloned members of the E2F family demonstrate that the E2F-E4-DNA complex likely involves at least six polypeptides including two E2F/DP1 heterodimers and two E4 proteins (CRESS and NEVINS 1994; OBERT et al. 1994; O'CONNOR and HEARING 1994; FAGAN et al. 1994; HELIN and HARLOW 1994; BANDARA et al. 1994), as schematically depicted in Fig. 5. Several lines of evidence are consistent with the 19-kDa E4 protein functioning as a homodimer (CRESS and NEVINS 1994; OBERT et al. 1994) and yeast two-hybrid analysis demonstrates that E4 can directly interact with E2F1 (FAGAN et al. 1994) and DP1 (CRESS and NEVINS 1994). The action of the E4 protein to generate the stable complex has been shown to require both the E2F1 and the DP1 components of the heterodimer (BANDARA et al. 1994; O'CONNOR and HEARING 1994; HELIN and HARLOW 1994). Thus, the DNA-protein complex at the E2 promoter likely contains an E4 dimer at its center tethering two heterodimers of E2F1/DP1 which in turn bind to the two inverted repeats in the E2 promoter.

The interactions of E2F with the adenovirus E4 protein and Rb appear to be mutually exclusive (O'CONNOR and HEARING 1994; FAGAN et al. 1994). This may result from the two proteins interacting with overlapping regions of E2F as suggested by a detailed mutagenesis study of the E2F1 and DP1 cDNAs (O'CONNOR and HEARING 1994). It is also clear that phosphorylation of E2F may play a role in modulating the interactions with E4 and Rb. Phosphorylation of serine residues of E2F1 at positions 332 and 337 blocks the interaction with the Rb protein while enhancing the interaction with E4 (FAGAN et al. 1994). Furthermore, two independent studies have found this region of E2F1 to be important in E2F–E4 interactions (HELIN and HARLOW 1994; O'CONNOR and HEARING 1994). Although it is presently unknown how the phosphorylation of E2F1 blocks the interaction with Rb, it is perhaps noteworthy that this region of E2F1 lies within a domain which is conserved among E2F1 homologs and has been termed the "marked box." This region of E2F1 appears to be important in E2F-mediated DNA bending, which is dramatically affected by interaction with Rb (HUBER et al. 1994; W.D. Cress and J.R. Nevins, unpublished data). Thus, it is possible that the marked box region controls the conformation of E2F changing its interaction with DNA and maximizing it for interactions with Rb versus adenovirus E4.

6 The Relationship Between Activation of E2F and Inactivation of p53 Function

Why do the DNA tumor viruses target both Rb and p53? Various experiments suggest that E1A expression is sufficient to drive quiescent cells into S phase, consistent with the additional studies that demonstrate that E2F1 overexpression can drive cells into S phase (KOWALIK et al. 1995; JOHNSON et al. 1993; WU and LEVINE 1994; QIN et al. 1994). Thus, if the strategy is to produce an S phase environment suitable for viral DNA replication, one might expect that this event alone would be sufficient. However, there is a strong correlation between the inactivation of both Rb and p53 in the course of DNA tumor virus infections as well as in human oncogenesis. This correlation is made particularly clear by examining the relationship between HPV gene expression and the state of Rb and p53 in cell lines derived from cervical carcinomas. The majority of human cervical carcinomas are associated with high risk HPV serotypes (ZUR HAUSEN et al. 1984). Howley and colleagues have shown that Rb and p53 remain wildtype in cervical carcinoma cell lines that expressed the HPV E6 and E7 proteins. In contrast, HPV-negative cervical carcinomas were found to possess inactivating mutations in both Rb and p53 (SCHEFFNER et al. 1991).

One possibility is that DNA tumor viruses eliminate p53 function that would suppress cell growth as a consequence of the activation of p21 expression, a potent inhibitor of G_1 cyclin kinase activity (HARPER et al. 1993; XIONG et al. 1993). A second, but not exclusive, possibility is that DNA tumor viruses target p53 to prevent loss of cell viability as a result of apoptosis. Degradation of viral DNA was first observed in cells infected by adenovirus E1B mutants (WHITE et al. 1984; PILDER et al. 1984). This apoptosis and associated DNA degradation is dependent on E1A expression, but also requires induction of p53, since there is no such response in cells that are lacking p53. Expression of the adenovirus E1B gene suppresses E1A-induced DNA degradation. As with E1A, the overexpression of the E2F1 transcription factor (SHAN and LEE 1994; KOWALIK et al. 1995; WU and LEVINE 1994; QIN et al. 1994) which drives quiescent cells into S phase also leads to an induction of apoptosis. When quiescent REF52 cells are infected with an adenovirus expressing E2F1 (and not E1A or E1B) cellular DNA synthesis is induced, but it is not complete since such cells never reach a G_2 DNA content.

The mechanism by which E2F1 overexpression drives apoptosis is not known. It is not unlikely, however, that induction of S phase by "activation" of E2F transcriptional activity represents only a partial signal for cell proliferation. This may be because E2F overexpression can activate only a subset of cellular promoters normally activated as quiescent cells enter S phase (J. DeGregori et al., submitted). This "partial" signal is detected, in large part dependent on p53, and apoptosis results. The DNA tumor viruses may have thus targeted p53 to delay DNA degradation induced as the host cell undergoes apoptosis.

7 Evolution of Common Strategies of the DNA Tumor Viruses

Although the inactivation of Rb and p53 by the DNA tumor virus oncoproteins is generally considered in the context of oncogenic transformation, these events must be important for the normal process of a productive infection by these viruses since it is the ability to replicate that defines the essential aspect of these viruses. Clearly these evolutionarily distinct viruses share a common need and target common cellular activities, not to transform cells but to replicate.

As discussed above, the DNA tumor viruses do have a common need to induce a quiescent, nondividing cell to enter S phase so as to create an environment that is favorable for viral DNA replication. The inactivation of Rb function through the action of E1A, T antigen, or E7 appears to facilitate this process by liberating the E2F transcription factor from inhibiting Rb complexes, thus leading to an induction of various genes that create the environment for DNA replication.

The viral-mediated inactivation of p53 function also appears to facilitate entry to S phase. These viruses likely target p53 since the expression of p53 can result in a G_1 arrest, particularly in response to DNA damaging events (LIVINGSTONE et al. 1992). Likewise, p53 can initiate a pathway of programmed cell death in response to various proliferative signals including the expression of E1A (WHITE 1993). Indeed, there is an induction of p53 expression in cells expressing E1A. The E1B 55-kDa protein, SV40 T antigen, and HPV E6 all block the action of p53 and thus block the apoptosis pathway. Thus, one might view these actions as "allowing" the E1A-mediated process of S phase induction to continue.

Given the common activities exhibited by adenovirus, SV40, and HPV, it is perhaps equally striking to find a distinct activity that is unique to polyomavirus, the middle T antigen mediated induction of tyrosine kinase activity. Presumably, a mechanism has evolved in polyomavirus to accomplish the same and result, the creation of a favorable environment for viral replication, without the need of eliminating the p53 suppression events. Polyoma large T antigen does target Rb and given the pairwise relationship between Rb inactivation and p53 inactivation seen with the other viral oncoproteins, one wonders whether the ultimate action of middle T might lead to the same end result as the other DNA tumor virus proteins that inactivate p53.

If the small DNA tumor viruses have a common need to drive quiescent cells into S phase, one might also anticipate that other DNA viruses that must replicate DNA at a high level, such as the herpesvirus or the poxviruses would also find this to be advantageous. Yet, there is no compelling evidence that any of the herpesviruses or poxviruses encode proteins that inactivate Rb or p53. It is striking, however, that many of the genes whose products create the S phase environment are found within the genomes of the large viruses of the herpesvirus and poxvirus family (ALBRECHT et al. 1992). Although the entire complement of S phase genes is not found within every virus of these groups, each virus does contain a ribonucleotide reductase, the rate-limiting enzyme in deoxynucleotide biosynthesis (THELANDER and REICHARD 1979). Moreover, a herpes simplex virus ribonucleotide reductase mutant is severely impaired for growth in vivo, both for growth in the eye and the trigerminal ganglion as well as reactivation from a latent infection (JACOBSON et al. 1989). Thus, a common need of the DNA viruses, whether oncogenic or not, may be to induce enzymatic activities that create an environment for viral DNA replication to take place in an efficient manner. If this need mechanistically involves the disruption of normal cell growth control events, transformation can result if the infection does not go to completion.

References

Albrecht JC, Nicholas J, Biller D, Cameron KR, Biesinger B, Newman C, Wittmann S, Craxton MA, Coleman H, Fleckenstein B. et al. (1992) Primary structure of the herpesvirus saimiri genome. J Virol 66: 5047–5058

Bagchi S, Raychaudhuri P, Nevins JR (1990) Adenovirus E1A proteins can dissociate heteromeric complexes involving the E2F transcription factor: a novel mechanism for E1A trans-activation. Cell 62: 659–669

Bagchi S, Weinmann R, Raychaudhuri P (1991) The retinoblastoma protein copurifies with E2F-I, an E1A-regulated inhibitor of the transcription factor E2F. Cell 65: 1063–1072

Bandara LR, La Thangue NB (1991) Adenovirus E1 a prevents the retinoblastoma gene product from complexing with a cellular transcription factor. Nature 351: 494–497

Bandara LR, Adamczewski JP, Hunt T, La Thangue NB (1990) Cyclin A and the retinoblastoma gene product complex with a common transcription factor. Nature 352: 249–251

Bandara LR, Lam EW-F, Sorensen TS, Zamanian M, Girling R, La Thangue NB (1994) DP-1: a cell cycle-regulated and phosphorylated component of transcription factor DRTF1/E2F which is functionally important for recognition by pRb and the adenovirus E4 orf 6/7 protein. EMBO 13: 3104–3114

Bjorklund S, Skog S, Tribukait B, Thelander L (1990) S-phase-specific expression of mammalian ribonucleotide reductase R1 and R2 subunit mRNAs. Biochemistry 29: 5452–5458

Cao L, Faha B, Dembski M, Tsai LH, Harlow E, Dyson N (1992) Independent binding of the retinoblastoma protein and p107 to the transcription factor E2F. Nature 355: 176–179

Chellappan SP, Hiebert S, Mudryj M, Horowitz JM, Nevins JR (1991) The E2F transcription factor is a cellular target for the RB protein. Cell 65: 1053–1061

Chellappan S, Kraus VB, Kroger B, Munger K, Howley PM, Phelps WC, Nevins JR (1992) Adenovirus E1A, simian virus 40 tumor antigen, and human papillomavirus E7 protein share the capacity to disrupt the interaction between transcription factor E2F and the retinoblastoma gene product. Proc Natl Acad Sci USA 89: 4549–4553

Chittenden T, Livingston DM, Kaelin WG Jr (1991) The T/E1A-binding domain of the retinoblastoma product can interact selectively with a sequence-specific DNA-binding protein. Cell 65: 1073–1082

Corbeil HB, Branton PE (1994) Functional importance of complex formation between the retinoblastoma tumor suppressor family and adenovirus E1A proteins as determined by mutational analysis of E1A conserved region 2. J Virol 68: 6697–6709

Cress WD, Nevins JR (1994) Interacting domains of E2F1, DP1, and the adenovirus E4 protein. J Virol 68: 4212–4219

Cress WD, Johnson DG, Nevins JR (1993) A genetic analysis of the E2F1 gene distinguishes regulation by Rb, p107, and adenovirus E4. Mol Cell Biol 13: 6314–6325

Dalton S (1992) Cell cycle regulation of the human cdc2 gene. EMBO J 11: 1797–1804

DeCaprio JA, Ludlow JW, Figge J, Shew JY, Huang CM, Lee WH, Marsilio E, Paucha E, Livingston DM (1988) SV40 large tumor antigen forms a specific complex with the product of the retinoblastoma susceptibility gene. Cell 54: 275–283

Defeo-Jones D, Huang PS, Jones RE, Haskell KM, Vuocolo GA, Hanobik MG, Huber HE, Oliff A (1991) Cloning of cDNAs for cellular proteins that bind to the retinoblastoma gene product. Nature 352: 251–254

DeGregori J, Kowalik T, Nevins JR (1995) Cellular targets for activation by the E2F1 transcription factor include DNA synthesis and G₁/S regulatory genes. Mol Cell Biol 15: 4215–4224

Devoto SH, Mudryj M, Pines J, Hunter T, Nevins JR (1992) A cyclin A-protein kinase complex possesses sequence-specific DNA binding activity: p33cdk2 is a component of the E2F-cylin A complex. Cell 68: 167–176

Dowdy SF, Hinds PW, Louie K, Reed SK, Arnold A, Weinberg RA (1993) Physical interaction of the retinoblastoma protein with human D cyclins. Cell 73: 499–511

Dulbecco R, Hartwell LH, Vogt M (1965) Induction of cellular DNA synthesis by polyoma virus. Proc Natl Acad Sci USA 53: 403

Dyson N, Howley PM, Munger K, Harlow E (1989) The human papilloma virus-16 E7 oncoprotein is able to bind to the retinoblastoma gene product. Science 243: 934–937

Dyson N, Guida P, McCall C, Harlow E (1992) Adenovirus E1A makes two distinct contacts with the retinoblastoma protein. J Virol 66: 4606–4611

Eckner R, Ewen ME, Newsome D, Gerdes M, DeCaprio JA, Lawrence JB, Livingston DM (1994) Molecular cloning and functional analysis of the adenovirus E1A-associated 300-kD protein (p300) reveals a protein with properties of a transcriptional adaptor. Genes Dev 8: 869–884

Engstrom Y, Eriksson S, Jildevik I, Skog S, Thelander L, Tribukait B (1985) Cell cycle-dependent expression of mammalian ribonucleotide reductase. Differential regulation of the two subunits. J Biol Chem 260: 9114–9116

Ewen ME, Xing YG, Lawrence JB, Livingston DM (1991) Molecular cloning, chromosomal mapping, and expression of the cDNA for p107, a retinoblastoma gene product-related protein. Cell 66: 1155–1164

Fagan R, Flint KJ, Jones N (1994) Phosphorylation of E2F-1 modulates its interaction with the retinoblastoma gene product and the adenovirus E4 19 kDa protein. Cell 78: 799–811

Fattaey AR, Harlow E, Helin K (1993) Independent regions of adenovirus E1A are required for binding to and dissociation of E2F-protein complexes. Mol Cell Biol 13: 7267–7277

Figge J, Webster T, Smith TF, Paucha E (1988) Prediction of similar transforming regions in simian virus 40 large T, adenovirus E1A, and myc oncoproteins. J Virol 62: 1814–1818

Flemington EK, Speck SH, Kaelin WG Jr (1993) E2F-1-mediated transactivation is inhibited by complex formation with the retinoblastoma susceptibility gene product. Proc Natl Acad Sci USA 90: 6914–6918

Frearson PM, Kit S, Dubbs DR (1965) Deoxythymidylate synthetase and deoxythymidine kinase activities of virus-infected animal cells. Cancer Res 25: 737

Frearson PM, Kit S, Dubbs DR (1966) Induction of dehydrofolate reductase activity by SV40 and polyoma virus. Cancer Res 26: 1653

Ginsberg D, Vairo G, Chittenden T, Xiao Z, Xu G, Wydner KL, DeCaprio JA, Lawrence JB, Livingston D (1994) E2F-4, a new member of the E2F transcription factor family, interacts with p107. Genes Dev 8: 2665–2679

Girling R, Partridge JF, Bandara LR, Burden N, Totty NF, Hsuan JJ, La Thangue NB (1993) A new component of the transcription factor DRTF1/E2F. Nature 362: 83–87

Hagemeier C, Cook A, Kouzarides T (1993) The retinoblastoma protein binds E2F residues required for activation in vivo and TBP binding in vitro. Nucleic Acids Res 21: 4998–5004

Hannon GJ, Demetrick D, Beach D (1993) Isolation of the Rb-related p130 through its interaction with cdk2 and cyclins. Genes Dev 7: 2378–2391

Hardy S, Shenk T (1989) E2F from adenovirus-infected cells binds cooperatively to DNA containing two properly oriented and spaced recognition sites. Mol Cell Biol 9: 4495–4506

Hardy S, Engel DA, Shenk T (1989) An adenovirus early region 4 gene product is required for induction of the infection-specific form of cellular E2F activity. Genes Dev 3: 1062–1074

Harlow E, Whyte P, Franza BR Jr, Schley C (1986) Association of adenovirus early-region 1A protein with cellular polypeptides. Mol Cell Biol 6: 1579–1589

Harper JW, Adami GR, Wei N, Keyomarsi K, Elledge SJ (1993) The p21 cdk-interacting protein Cip1 is a potent inhibitor of G₁ cyclin-dependent kinases. Cell 75: 805–816

Hartwell L, Vogt M, Dulbecco R (1965) Induction of cellular DNA synthesis by polyoma. II. Increase in the rate of enzyme synthesis after infection with polyoma virus in mouse embryo kidney cells. Virology 27: 262

Hatanaka M, Dulbecco R (1966) Induction of DNA synthesis by SV40. Proc Natl Acad Sci USA 56: 736–740

Helin K, Harlow E (1994) Heterodimerization of the transcription factors E2F-1 and DP-1 is required for binding to the adenovirus E4 (ORF6/7) protein. J Virol 68: 5027–5035

Helin K, Lees JA, Dyson N, Harlow E, Fattaey A (1992) A cDNA encoding a pRB-binding protein with properties of the transcription factor E2F. Cell 70: 337–350

Helin K, Harlow E, Fattaey A (1993a) Inhibition of E2F-1 transactivation by direct binding of the retinoblastoma protein. Mol Cell Biol 13: 6501–6508

Helin K, Wu C-L, Fattaey AR, Lees JA, Dynlacht BD, Ngwu C, Harlow E (1993b) Heterodimerization of the transcription factors E2F-1 and DP-1 leads to cooperative *trans*-activation. Genes Dev 7: 1850–1861

Hiebert SW (1993) Regions of the retinoblastoma gene product required for its interaction with the E2F transcription factor are necessary for E2 promoter repression and pRb-mediated growth suppression. Mol Cell Biol 13: 3384–3391

Hiebert SW, Lipp M, Nevins JR (1989) E1A-dependent *trans*-activation of the human MYC promoter is mediated by the E2F factor. Proc Natl Acad Sci USA 86: 3594–3598

Hiebert SW, Blake M, Azizkhan J, Nevins JR (1991) Role of E2F transcription factor in E1A-mediated *trans*-activation of cellular genes. J Virol 65: 3547–3552

Hiebert SW, Chellappan SP, Horowitz JM, Nevins JR (1992) The interaction of RB with E2F coincides with an inhibition of the transcriptional activity of E2F. Genes Dev 6: 177–185

Huang MM, Hearing P (1989) The adenovirus early region 4 open reading frame 6/7 protein regulates the DNA binding activity of the cellular transcription factor, E2F, through a direct complex. Genes Dev 3: 1699–1710

Huang PS, Patrick DR, Edwards G, Goodhart PJ, Huber HE, Miles L, Garsky VM, Oliff A, Heimbrook DC (1993) Protein domains governing interactions between E2F, the retinoblastoma gene product, and human papillomavirus type 16 E7 protein. Mol Cell Biol 13: 953–960

Huber HE, Goodhart PJ, Huang PS (1994) Retinoblastoma protein reverses DNA bending by transcription factor E2F. J Biol Chem 269: 6999–7005

Ikeda M-A, Nevins JR (1993) Identification of distinct roles for separate E1A domains in the disruption of E2F complexes. Mol Cell Biol 13: 7029–7035

Jacobson JG, Leib DA, Goldstein DJ, Bogard CL, Schaffer PA, Weller SK, Coen DM (1989) A herpes simplex virus ribonucleotide reductase deletion mutant is defective for productive acute and reactivatable latent infections of mice and for replication in mouse cells. Virology 173: 276–283

Johnson DG, Schwarz JK, Cress WD, Nevins JR (1993) Expression of transcription factor E2F1 induces quiescent cells to enter S phase. Nature 365: 349–352

Johnson DG, Cress WD, Jakoi L, Nevins JR (1994) Oncogenic capacity of the E2F1 gene. Proc Natl Acad Sci USA 91: 12823–12827

Kaelin WG, Krek W, Sellers WR, DeCaprio JA, Ajchenbaum F, Fuchs CS, Chittenden T, Li Y, Farnham PJ, Blanar MA et al (1992) Expression cloning of a cDNA encoding a retinoblastoma-binding protein with E2F-like properties. Cell 70: 351–364

Kara J, Weil R (1967) Specific activation of the DNA synthesizing apparatus in contact inhibited cells by polyoma virus. Proc Natl Acad Sci USA 57: 63

Kato J, Matsushime H, Hiebert SW, Ewen ME, Sherr CJ (1993) Direct binding of cyclin D to the retinoblastoma gene product (pRb) and pRb phosphorylation by the cyclin D-dependent kinase CDK4. Genes Dev 7: 331–342

Kit S, Dubbs DR, Frearson PM (1966a) Enzymes of nucleic acid metabolism in cells infected with polyoma virus. Cancer Res 26: 638

Kit S, Dubbs DR, Frearson PM, Melnick JL (1966b) Enzyme induction in SV40-infected green monkey kidney cultures. Virology 29: 69

Kit S, De Torres RA, Dubbs DR, Salvi ML (1967a) Induction of cellular deoxyribonucleic acid synthesis by simian virus 40. J Virol 1: 738

Kit S, Piekarski LJ, Dubbs DR (1967b) DNA polymerase induced by simian virus 40. J Gen Virol 1: 163

Kovesdi I, Reichel R, Nevins JR (1986a) E1A transcription induction: enhanced binding of a factor to upstream promoter sequences. Science 231: 719–722

Kovesdi I, Reichel R, Nevins JR (1986b) Identification of a cellular transcription factor involved in E1A *trans*-activation. Cell 45: 219–228

Kowalik TF, DeGregori J, Schewarz JK, Nevins JK (1995) E2F1 Overexpression in quiescent fibroblasts leads to induction of celluar DNA synthesis and apoptosis. J Virol (in press)

Lam EW, Watson RJ (1993) An E2F-binding site mediates cell-cycle regulated repression of mouse B-myb transcription. EMBO J 12: 2705–2713

Ledinko N (1968) Enhanced deoxyribonucleic acid polymerase activity in human embryonic kidney cultures infected with adenovirus 2 and 12. J Virol 2: 89–98

Li Y, Graham C, Lacy S, Duncan AMV, Whyte P (1993) The adenovirus E1A-associated 130-kd protein is encoded by a member of the retinoblastoma gene family and physically interacts with cyclins A and E. Genes Dev 7: 2366–2377

Livingstone LR, White A, Sprouse J, Livanos E, Jacks T, Tlsty TD (1992) Altered cell cycle arrest and gene amplification potential accompany loss of wild-type p53. Cell 70: 923–935

Loeken MR, Brady J (1989) Analysis of regulatory sequences and changes in binding activity of ATF and EIIF following adenovirus infection. J Biol Chem 264: 6572–6579

Marton MJ, Baim SB, Ornelles DA, Shenk T (1990) The adenovirus E4 17-kilodalton protein complexes with the cellular transcription factor E2F, altering its DNA-binding properties and stimulating E1A-independent accumulation of E2 mRNA. J Virol 64: 2345–2359

Matsushime H, Roussel MF, Ashmun RA, Sherr CJ (1991) Colony-stimulating factor 1 regulates novel cyclins during the G1 phase of the cell cycle. Cell 65: 701–713

Mayol X, Grana X, Baldi A, Sang N, Hu Q, Giordano A (1993) Cloning of a new member of the retinoblastoma gene family (pRb2) which binds to the E1A transforming domain. Oncogene 8: 2561–2566

Neill SD, Nevins JR (1991) Genetic analysis of the adenovirus E4 6/7 trans-activator: interaction with E2F and induction of a stable DNA-protein complex are critical for activity. J Virol 65: 5364–5373

Neill SD, Hemstrom C, Virtanen A, Nevins JR (1990) An adenovirus E4 gene product trans-activates E2 transcription and stimulates stable E2F binding through a direct association with E2F. Proc Natl Acad Sci USA 87: 2008–2012

Nevins JR (1987) Regulation of early adenovirus gene expression. Microbiol Rev 51: 419–430

Obert S, O'Connor RJ, Schmid S, Hearing P (1994) The adenovirus E4-6/7 protein transactivates the E2 promoter by inducing dimerization of a heteromeric E2F complex. Mol Cell Biol 14: 1333–1346

O'Connor RJ, Hearing P (1991) The C-terminal 70 amino acids of the adenovirus E4-ORF6/7 protein are essential and sufficient for E2F complex formation. Nucleic Acids Res 19: 6579–6586

O'Connor RJ, Hearing P (1994) Mutually exclusive interaction of the adenovirus E4-6/7 protein and the retinoblastoma gene product with internal domains of E2F-1 and DP-1. J Virol 68: 6848–6862

Pilder S, Logan J, Sheck T (1984) Deletion of the gene encoding the adenovirus 5 early region 1b 21,000-molecular-weight polypeptide leads to degradation of viral and host cell DNA. J Virol 52: 664–671

Qin X-Q, Livingston DM, Kaelin WG, Adams PD (1994) Deregulated transcription factor E2F-1 expression leads to S-phase entry and p53-mediated apoptosis. Proc Natl Acad Sci USA 91: 10918–10922

Raychaudhuri P, Bagchi S, Neill SD, Nevins JR (1990) Activation of the E2F transcription factor in adenovirus-infected cells involves E1A-dependent stimulation of DNA-binding activity and induction of cooperative binding mediated by an E4 gene product. J Virol 64: 2701–2710

Raychaudhuri P, Bagchi S, Devoto SH, Kraus VB, Moran E, Nevins JR (1991) Domains of the adenovirus E1A protein required for oncogenic activity are also required for dissociation of E2F transcription factor complexes. Genes Dev 5: 1200–1211

Reichel R, Neill SD, Kovesdi I, Simon MC, Raychaudhuri P, Nevins JR (1989) The adenovirus E4 gene, in addition to the E1A gene, is important for trans-activation of E2 transcription and for E2F activation. J Virol 63: 3643–3650

Scheffner M, Munger K, Byrne JC, Howley PM (1991) The state of the p53 and retinoblastoma genes in human cervical carcinoma cell lines. Proc Natl Acad Sci USA 88: 5523–5527

Schwarz JK, Bassing CH, Kovesdi I, Datto MB, Blazing M, George S, Wang X, Nevins JR (1995) Expression of the E2F1 transcription factor overcomes type β transforming growth factor-mediated growth suppression. Proc Natl Acad Sci USA 92: 483–487

Shan B, Lee W (1994) Deregulated expression of E2F-1 induces S-phase entry and leads to apoptosis. Mol Cell Biol 14: 8166–8173

Shan B, Zhu X, Chen PL, Durfee T, Yang Y, Sharp D, Lee WH (1992) Molecular cloning of cellular genes encoding retinoblastoma-associated proteins: identification of a gene with properties of the transcription factor E2F. Mol Cell Biol 12: 5620–5631

Sheinin R (1966) Studies on the thymidine kinase activity of mouse embryo cells infected with polyoma virus. Virology 28: 47

Shirodkar S, Ewen M, DeCaprio JA, Morgan J, Livingston DM, Chittenden T (1992) The transcription factor E2F interacts with the retinoblastoma product and a p107-cyclin A complex in a cell cycle-regulated manner. Cell 68: 157–166

Singh P, Wong SW, Hong W (1994) Overexpression of E2F-1 in rat embryo fibroblasts leads to neoplastic transformation. EMBO J 13: 3329–3338

Slansky JE, Li Y, Kaelin WG, Farnham PJ (1993) A protein synthesis-dependent increase in E2F1 mRNA correlates with growth regulation of the dihydrofolate reductase promoter. Mol Cell Biol 13: 1610–1618

Thalmeier K, Synovzik H, Mertz R, Winnacker EL, Lipp M (1989) Nuclear factor E2F mediates basic transcription and trans-activation by Ela of the human MYC promoter. Genes Dev 3: 527–536

Thelander L, Reichard P (1979) Reduction of ribonucleotides. Annu Rev Biochem 48: 133–158

Tsai LH, Harlow E, Meyerson M (1991) Isolation of the human cdk2 gene that encodes the cyclin A- and adenovirus E1A-associated p33 kinase. Nature 353: 174–177

Wang HH, Rikitake Y, Carter MC, Yaciuk P, Abraham SE, Zerler B, Moran E (1993) Identification of specific adenovirus E1A N-terminal residues critical to the binding of cellular proteins and to the control of cell growth. J Virol 67: 476–488

White E (1993) Death-defying acts: a meeting review on apoptosis. Genes Dev 7: 2277–2284

White E, Grodzicker T, Stillman BW (1984) Mutations in the gene encoding the adenovirus early region 1B 19,000-molecular-weight tumor antigen cause the degradation of chromosomal DNA. J Virol 52: 410

Whyte P, Buchkovich KJ, Horowitz JM, Friend SH, Raybuck M, Weinberg RA, Harlow E (1988) Association between an oncogene and an anti-oncogene: the adenovirus E1A proteins bind to the retinoblastoma gene product. Nature 334: 124–129

Whyte P, Williamson NM, Harlow E (1989) Cellular targets for transformation by the adenovirus E1A proteins. Cell 56: 67–75

Wu L, Berk AJ (1988) Transcriptional activation by the pseudorabies virus immediate early protein requires the TATA box element in the adenovirus 2 E1B promoter. Virology 167: 318–322

Wu X, Levine AJ (1994) p53 and E2F-1 cooperate to mediate apoptosis. Proc Natl Acad Sci USA 91: 3602–3606

Xiong Y, Connolly T, Futcher B, Beach D (1991) Human D type cyclin. Cell 65: 691–699

Xiong Y, Hannon GJ, Zhang H, Casso D, Kobayashi R, Beach D (1993) p21 is a universal inhibitor of cyclin kinases. Nature 366: 701–704

Yamashita T, Shimojo H (1969) Induction of cellular DNA synthesis by adenovirus 12 in human embryo kidney cells. Virology 38: 351–355

Yee S, Branton PE (1985) Detection of cellular proteins associated with human adenovirus type 5 early 1A polypeptides. Virology 147: 142–153

Yee AS, Raychaudhuri P, Jakoi L, Nevins JR (1989) The adenovirus-inducible factor E2F stimulates transcription after specific DNA binding. Mol Cell Biol 9: 578–585

Zamanian M, La Thangue NB (1993) Transcriptional repression by the Rb-related protein p107. Mol Biol Cell 4: 389–396

Zhu L, van den Heuvel S, Helin K, Fattaey A, Ewen M, Livingston D, Dyson N, Harlow E (1993) Inhibition of cell proliferation by p107, a relative of the retinoblastoma protein. Genes Dev 7: 1111–1125

zur Hausen H, Gissmann L, Schlehofer JR (1984) Viruses in the etiology of human genital cancer. Prog Med Virol 30: 170–186

The Cellular Effects of E2F Overexpression

P.D. Adams and W.G. Kaelin, Jr.

1 Summary

The product of the retinoblastoma tumor-suppressor gene (*RB*) is a ubiquitously expressed, 105-kDa nuclear phosphoprotein (pRB). The pRB protein negatively regulates the cellular G_1/S phase transition, and it is at this point in the cell cycle that it is thought to play its role as a tumor suppressor. The growth-inhibitory effects of pRB are exerted, at least in part, through the E2F family of transcription factors. This chapter reviews the insights into the mechanism of action of the E2F family members that have been obtained through overexpression studies. Studies in $RB^{-/-}$ SAOS-2 cells have provided evidence in support of the hypothesis that the E2F family members are negatively regulated by pRB and the related protein p130. In particular, the results obtained are consistent with the earlier biochemical data which suggested that E2F1 is regulated primarily by pRB, and E2F4 by p130. Results relating to p107 are also discussed. Consistent with the proposed role of pRB and E2F1 as coregulators of entry into S phase, experiments have demonstrated that overexpression of E2F1 is sufficient to override the cell cycle arrests caused by serum deprivation of fibroblasts or transforming growth factor-β (TGFβ) treatment of mink lung epithelial cells. However, at least in the case of the serum deprivation induced arrest, the ultimate result of E2F1 overexpression is death by p53-dependent apoptosis. In

Dana-Farber Cancer Institute and Harvard Medical School, 44 Binney Street, Boston, MA 02115, USA light of this and other data, a model is discussed as to how functional inactivation of pRB and p53 might cooperate to promote tumorigenesis. A number of studies have demonstrated the oncogenic potential of E2F family members, at least under certain conditions. This is, again, in keeping with the notion that these proteins play a critical role in controlling cellular proliferation.

2 Introduction

As detailed in other chapters, E2F was originally identified as a DNA-binding activity that is stimulated by the adenovirus E1a protein (Kovesdi et al. 1986). This activity was shown to bind to sites within the adenovirus E2 promoter that are required for transcription of the E2 gene. Independent studies on the differentiation of F9 embryonal carcinoma cells also identified a closely related or identical differentiation-regulated transcription factor (DRTF; La Thangue and Rigby 1987). E2F DNA-binding activity in vivo is likely due to heterodimeric complexes containing an E2F family member bound to a DP family member (Bandara et al. 1993; Helin et al. 1993; Huber et al. 1993; Krek et al. 1993). At least five different E2F family members (E2F1–5) and at least three DP family members (DP1–3) have been cloned (Beijersbergen et al. 1994; R. Bernards 1995, personal communication; Ginsberg et al. 1994; Girling et al. 1994; Helin et al. 1992; Ivey-Hoyle et al. 1993; Kaelin et al. 1992; Lees et al. 1993; Sardet et al. 1995; Shan et al. 1992; Wu et al. 1995). This review uses E2F as a generic term to refer to the hererodimeric activities which bind to the consensus sequence TTT(G/C)(G/C)CG(G/C).

E2F transcriptional activity is regulated, at least in part, through association with "pocket proteins" such as the retinoblastoma susceptibility gene product, pRB, a known negative regulator of the G_1/S transition (Ewen 1994). pRB binds to certain E2F family members and in so doing inhibits their ability to activate transcription (Flemington et al. 1993; Helin et al. 1993 and references therein). Furthermore, in certain settings, pRB/E2F complexes may, in a DNA-binding dependent manner, actively repress transcription, perhaps by interacting with adjacent, non-E2F transcription factors (Sellers et al. 1995; Qin et al. 1995). Ela-mediated transformation is linked to its ability to disrupt E2F/pRB complexes, thereby liberating "free" E2F which in turn leads to an increase in the transcription of certain E2F-dependent genes (Hamel et al. 1992; Hiebert et al. 1992; Weintraub et al. 1992; Zamanian and La Thangue 1992). Transformation by the transforming oncoproteins of simian virus 40 (SV40) and human papilloma virus (HPV), large T antigen and E7, respectively, also requires disruption of cellular E2F/pRB complexes, suggesting that deregulated E2F activity is a necessary event for transformation (Moran 1993). Furthermore, all stable, naturally occurring, inactivating pRB mutations map to the T, Ela, and E7 binding region or "pocket", a region that is also implicated in E2F binding

(EWEN 1994). Other "pocket proteins" known to associate with E2Fs and likely to regulate E2F activity include p107 and p130 (COBRINIK et al. 1993; EWEN 1994; HANNON et al. 1993). E2F activity is likely also regulated through posttranslational modifications of the E2F and DP components and, at least in the case of E2F-1, through cell cycle dependent changes in transcription (DYNLACHT et al. 1994; HSIAO et al. 1994; JOHNSON et al. 1994; KAELIN et al. 1992; KREK et al. 1994; LI et al. 1993; NEUMAN et al. 1994; PEEPER et al. 1995; XU et al. 1994). Analysis of the promoters of a number cellular genes implicated in the control of cell cycle progression reveals that they contain E2F-binding sites. These include the proto-oncogenes c-*myc* and B-*myb*, genes encoding enzymes required for DNA synthesis such as DHFR, thymidine kinase, thymidylate synthase and DNA polymerase α, and genes encoding components of the basic cell cycle clock such as cdc-2, cyclin A and cyclin D1 (HERBER et al. 1994; JOLIFF et al. 1991; LI et al. 1993; NEVINS 1992; OSWALD et al. 1994; PEARSON et al. 1991; PHILIPP et al. 1994; ROUSSEL et al. 1994; YAMAMOTO et al. 1994).

The knowledge that E2F sites are present in a number of genes implicated in cell growth control, that E2F is negatively regulated by the product of a known tumor suppressor gene (*RB*), and that this regulation is disrupted during the course of viral transformation and by naturally occuring *RB* gene mutations, taken together, strongly suggests that E2F is likely a key regulator of cell cycle progression. If so, it might be expected that under particular circumstances over-expression or mutation of E2Fs promotes cellular transformation. This chapter reviews the insights into role of E2F in the cell cycle obtained from overexpression studies. Such studies are potentially of interest because they can address the extent to which particular E2F family members are sufficient for a particular cell cycle transition. An attempt is made to view the data in light of other biochemical properties of E2F.

3 Studies in SAOS-2 Cells

A number of studies investigating the cellular consequences of E2F overexpression have been performed in RB$^{-/-}$ SAOS-2 cells. Studies in these cells have provided support for the hypothesis that E2F1 is a downstream target of pRB, and that the regulatory role of these two proteins is played out at the G_1/S transition. Reintroduction of wild-type pRB into SAOS-2 cells results in a late G_1 cell cycle arrest and formation of morphologically distinct "large cells" (HINDS et al. 1992; HUANG et al. 1988; MITTNACHT and WEINBERG 1991; QIAN et al. 1992; QIN et al. 1992). Cotransfection of wild-type E2F1 with pRB into these cells inhibits the ability of pRB to induce both the late G_1 arrest and formation of "large cells" (QIN et al. 1995; ZHU et al. 1993). Mutational analysis of E2F1 supports the notion that override of the pRB induced G_1/S arrest is by virtue of E2F1's ability to act as a transcription factor. Firstly, an E2F1 mutant that is unable to bind pRB, but which retains the ability to transactivate, is likewise able to overcome a pRB arrest (QIN et al. 1995). Secondly, E2F1 mutants which bind to pRB but not to DNA are unable to overcome a pRB arrest, suggesting that E2F1 does not merely behave like T antigen or E1a in these assays and displace other effectors from the pRB pocket (QIN et al. 1995). The ability of E2F1 to overcome a pRB arrest is not associated with overt pRB phosphorylation, and, indeed, E2F1 can overcome the action of a nonphosphorylatable and hence constitutively active pRB mutant (QIN et al. 1995). These observations are consistent with free E2F1 functioning downstream of pRB phosphorylation. Of note, an E2F1 mutant (truncated at amino acid 196) which retains the ability to bind to canonical E2F sites, but which lacks the E2F1 transactivation/pRB binding domain, prevents the induction of "large cells" by pRB but does not overcome the ability of pRB to induce a G_1/S block. The latter requires an E2F1 species with an intact DNA binding domain and a functional transactivation domain (QIN et al. 1995). These findings are consistent with a model in which E2F/pRB complexes bind to, and actively repress the transcription of, certain E2F responsive promoters, rather than merely representing "inactivated" or "sequestered" E2F. According to this model, "large cell" morphology depends upon repression of transcription by E2F/pRB and can be relieved by a dominant-negative acting E2F1 DNA-binding domain, whereas the progression into S phase requires both alleviation of transcriptional repression and transcriptional activation of certain E2F-dependent genes.

These results are in keeping with earlier studies which suggested that E2F sites, both in certain artificial promoters and in the cdc-2, B-myb and E2F1 promoters, mediate cell cycle dependent transcriptional repression in cells containing wild-type, but not mutant, pRB (DALTON 1992; HSIAO et al. 1994; JOHNSON et al. 1994; LAM and WATSON 1993; NEUMAN et al. 1994; WEINTRAUB et al. 1992). Thus, as illustrated in Fig. 1, at least three activity states of E2F-responsive genes can be envisioned; repressed by a pocket protein/E2F complex, activated by free E2F, and a basal state in which transcriptional activity is determined by the binding of non-E2F transcription factors. In support of such a model, an E2F1 mutant (truncated at amino acid 368) that is competent to heterodimerize with DP1 and bind to DNA but can no longer transactivate or bind to pRB is able to relieve pRB mediated transcriptional repression of a naturally occurring E2F responsive promoter in SAOS-2 cells, presumably by preventing the binding of pRB/E2F complexes to the promoter (QIN et al. 1995). Recently, a transcriptional repression domain has been identified within pRB (SELLERS et al. 1995). It is perhaps note-worthy that c-myc, the product of a suspected E2F target gene, can likewise overcome a pRB-induced G_1/S block, raising the possibility that the ability of E2F1 to overcome a pRB block is due at least in part to activation of c-myc (GOODRICH and LEE 1992). These experiments did not assess the ability of c-myc to override the formation of pRB induced "large cells."

Similarly to pRB, expression of p107 or p130 in SAOS-2 cells results in a G_1/S arrest (VAIRO et al. 1995; ZHU et al. 1993). The ability of p130 to induce growth arrest in these cells is overridden by E2F4 and to a lesser extent by E2F1 (VAIRO

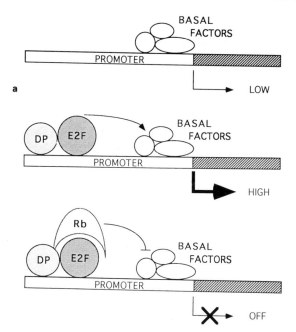

Fig. 1. E2F responsive genes have three activity states. **a** In the absence of "free" E2F and E2F/pRB complexes E2F responsive genes have basal activity due to the activity of non'-E2F transcription factors and the basal transcription apparatus. **b** In the presence of "free" E2F uncomplexed to pocket proteins the promoter activity is upregulated through transactivation by E2F. **c** In the presence of E2F/pRB complexes the promoter activity is repressed to a level below that in **a**. Thus, E2F/pRB complexes constitute active repression factors

et al. 1995). Conversely, E2F4 overrides a pRB growth arrest less efficiently than E2F1 (Vairo et al. 1995). The p107-induced growth arrest is not overridden by E2F1, and there is no published report of an override by E2F4 (Zhu et al. 1993). In vivo, E2F1 associates preferentially with pRB, and E2F4 with p107 and p130 (Beijersbergen et al. 1994; Ginsberg et al. 1994; Helin et al. 1992; Kaelin et al. 1992; Lees et al. 1993; Vairo et al. 1995). Thus, taken together these observations are consistent with E2F1 being regulated primarily by pRB, and E2F4 by p130 and perhaps p107. Recent studies suggest that p107 may bind to and repress transactivation by c-myc, suggesting another mechanism by which p107 might suppress cell growth (Beijersbergen et al. 1994; Gu et al. 1994). Indeed, c-myc is capable of overriding the p107-induced growth arrest in SAOS-2 cells (Beijersbergen et al. 1994). In addition to relieving the p130 growth arrest, E2F4, when transfected into SAOS-2 cells, interacts synergistically with its hetero-dimeric partner DP1 to promote progression into S phase (Beijersbergen et al. 1994). In summary, in RB$^{-/-}$ SAOS-2 osteosarcoma cells E2F1 and E2F4 override the growth-inhibitory effects of the pocket proteins pRB, p107, and p130 with the specificities that would be predicted, at least to a certain extent, from their respective associations in vivo. Do such observations extend to other cell types and other growth inhibitory stimuli?

4 Override of Cell Cycle Arrest
   Caused by Serum Withdrawal and TGFβ

Nevins and coworkers have shown that overexpression of E2F1 in REF 52 cells inhibits their ability to enter G_0 upon serum removal (JOHNSON et al. 1993). Similar results were obtained by using Rat1a fibroblasts stably overexpressing the E2F1 protein (P.D. Adams and W.G. Kaelin Jr., unpublished data). Futhermore, untimely production of the E2F1 protein is sufficient to stimulate quiescent fibroblasts to enter S phase (JOHNSON et al. 1993; QIN et al. 1994; SHAN and LEE 1994). In keeping with the structural requirements for E2F1 to override a pRB-induced G_1/S block in SAOS-2 cells (QIN et al. 1995), mutational analysis shows that the serum override effect of E2F1, as measured by entry into S phase, requires intact DNA binding and transactivation domains, suggesting that E2F1 is acting as a transcription factor in these assays (JOHNSON et al. 1993; SHAN and LEE 1994). Nevins and coworkers have additionally shown that overexpression of E2F1 in mink lung epithelial cells is able to override the middle G_1 growth arrest induced by TGFβ (SCHWARZ et al. 1995). This is consistent with the observation that viral onco-proteins that disrupt "pocket protein" complexes are able to block the TGFβ arrest (LAIHO et al. 1990). Recent data suggest that the ability of TGFβ to induce a cell cycle arrest is linked to its ability to inhibit the putative pRB kinases, cyclin D/cdk4 and cyclin E/cdk2. This inhibition is achieved through the suppression of transla-tion of cdk4 and the action of the cdk inhibitors, p27 and p15, on cyclin E/cdk2 and cyclin D/cdk4, respectively (EWEN et al. 1993, 1995; HANNON and BEACH 1994; POLYAAK et al. 1994). Presumably, overexpression of E2F1, a major effector of pRB, negates the requirements for a functional pRB kinase for the G_1/S transition. This is consistent with the lack of a requirement for D cyclins in RB$^{-/-}$ cells and the inability of the polypeptide cdk4 inhibitors p16 and p18 to arrest growth of RB$^{-/-}$ cells (BARTKOVA et al. 1994; GUAN et al. 1994; LUKAS et al. 1994; MULLER et al. 1994).

Thus, at least as measured by induction of DNA synthesis, E2F1 over-expression is sufficient to override the negative effects on cell growth resulting from TGFβ treatment of mink lung epithelial cells or serum deprivation of fibroblasts. At least under certain conditions, however, cells which are induced to enter S phase by untimely production of E2F1 proceed to undergo apoptosis.

5 Deregulated E2F1 Expression Ultimately Results
   in p53-Dependent Apoptosis

QIN et al. (1994) observed nuclear morphological changes indicative of apoptosis when a variety of cell types were transfected with plasmids encoding E2F1 and subsequently placed into low serum. This effect depends upon an intact DNA binding and transactivation domain. Two groups have been successful in generating stable rat fibroblast cell lines in which the human E2F1 gene is under the control of a regulatable promoter (QIN et al. 1994; SHAN and LEE 1994). When these cells are incubated in low serum and E2F1 production induced, the cells enter S phase, in keeping with the earlier microinjection data of JOHNSON et al.(1993). FACS analysis shows that the cells seemingly undergo a full round of DNA synthesis as they double their DNA content. Furthermore, the kinetics of entry into S phase are very similar to those observed after serum stimulation. However, about 20–30 h after the passage through S phase, the E2F1 expressing cells undergo death by apoptosis. In transient transfection experiments it was shown that E2F1-induced apoptosis in the presence of low serum is at least in part p53 dependent (QIN et al. 1994). Similarly, WU and LEVINE (1994) have shown, using a fibroblast cell line expressing a temperature-sensitive p53 mutant and over-expressing E2F1, that wild-type p53 and E2F1 cooperate to induce apoptosis.

The observation that deregulated E2F1 activity promotes apoptosis by a p53-dependent pathway may help to explain why DNA tumor viruses functionally inactivate both pRB and p53 (MORAN 1993). By this model, loss of pRB alone, and deregulation of E2F, promotes death by p53-dependent apoptosis. Consistent with this notion, RB$^{-/-}$ mice die between days 12 and 16 of development, showing signs of excessive apoptosis in peripheral and central neural tissues (CLARKE et al. 1992; JACKS et al. 1992; LEE et al. 1992). Using in vivo bromodeoxyuridine labeling, LEE and coworkers (1994) have demonstrated a spatial correlation between the tissue localization of cells undergoing apoptosis and cells undergoing aberrant DNA synthesis. Furthermore, many of the apoptotic cells can be labeled with BrdU, indicating that they have recently passed through S phase. A number of studies suggest that apoptosis occurring in vivo, as consequence of loss of or functional inactivation of pRB, is p53 dependent. Analysis of the developing ocular lens of the RB$^{-/-}$ mouse embryos reveals elevated numbers of cells undergoing apoptosis and DNA synthesis and a failure to express markers of differentiation (MORGENBESSER et al. 1994). However, apoptosis is largely suppressed in the lenses of RB$^{-/-}$/p53$^{-/-}$ mice. Interestingly, the p53 status does not affect the extent of aberrant DNA synthesis or expression of differentiation markers. This is consistent with a model in which p53 is required for apoptosis as a consequence of aberrant cell cycle progression.

Additional support for this model comes from studies of transgenic mice expressing the transforming oncoproteins of small DNA tumor viruses. Firstly, p53$^{+/+}$ mice expressing wild-type SV40 T antigen in B and T cells and the choroid plexus epithelium (CPE) develop tumors in those tissues. In contrast, p53$^{+/+}$ mice expressing a truncated T antigen that binds pRB, but not p53, develop only CPE tumors which develop slowly and show morphological signs of apoptosis (SAENZ ROBLES et al. 1994; SYMONDS et al. 1994). Expression of either wild-type or mutant T antigen in p53$^{-/-}$ mice induces full tumor formation with no signs of apoptosis (SYMONDS et al. 1994). Secondly, p53$^{+/+}$ mice expressing the HPV-16 E7 protein in photoreceptor cells exhibit retinal degeneration as a result of apoptosis, whereas expression of HPV-16 E7 in the same cells of p53$^{-/-}$ mice initiates retinal tumor formation (HOWES et al. 1994). Thirdly, transgenic expression of the HPV E7 protein in the mouse eye lens inhibits cellular differentiation and stimulates proliferation and apoptosis (PAN and GRIEP 1994). These effects are dependent upon the pRB/p107 binding function of E7. In doubly transgenic E6+E7 lenses the level of apoptosis is reduced relative to E7 alone (PAN and GRIEP 1994), consistent with it being a p53-dependent process. In cell culture systems expression of Ela in BRK or REF52 cells induces p53-dependent apoptosis (DEBBAS and WHITE 1993; LOWE 1993). Although the pRB binding CR2 domain of Ela is not absolutely necessary for this effect, mutational analysis shows that it does play a role. It seems likely that the pRB binding and p300 binding functions of Ela cooperate to induce S phase entry and apotosis (HOWE et al. 1990; MORAN and ZERLER 1988; MYMRYK et al. 1994; STEIN et al. 1990; WHITE et al. 1991).

In summary, much evidence suggests that functional inactivation of pRB leads to p53-dependent apoptosis. That loss of pRB promotes p53-dependent apoptosis might also be true in most human tissues, in view of the observation that p53 is frequently altered in RB$^{-/-}$ adult solid tumors. However, many human reinoblastomas contain wild-type p53 (HAMEL et al. 1993), suggesting that human retinoblasts, unlike their murine counterparts, may be relatively resistant to the killing effect of deregulated E2F (QIN et al. 1994; SHAN and LEE 1994; WU and LEVINE 1994).Such a difference between murine and human retinal cells might account for the observation that RB$^{-/+}$ humans, but not RB$^{-/+}$ mice, are predisposed to reinoblastoma development (CLARKE et al. 1992; JACKS et al. 1992; LEE et al. 1992).

6 Deregulated Expression or Mutation of E2Fs Can Promote Cellular Transformation

Other studies have demonstrated that at least under certain conditions deregulated expression of wild-type or mutant E2F family members can promote transformation. SINGH et al.(1994) showed that REF clonal cell lines stably over-expressing the E2F1 protein are transformed, as indicated by a number of criteria including the loss of contact inhibition, the ability to grow in soft agar and low serum and the ability to form tumors in nude mice. Transformation requires the DNA binding and transactivation domains of E2F1. The p53 status of the REFs used in this study was not investigated, although other studies have shown REFs to contain wild-type p53. If so, the ability of these cells to grow in low serum is in apparent contrast to the earlier studies, demonstrating p53-dependent E2F1-induced apoptosis in low serum (QIN et al. 1994; SHAN and LEE 1994). The cell lines used in the study by SINGH et al. had been through a drug selection process and the lines expanded twice from low density. Thus, these cells might have accumulated other mutations, such as inactivation of p53, during this process. Another study has demonstrated that E2F1 cooperates with DP1 and activated ras to transform primary REFs, as assayed by growth in soft agar and tumorigenicity in nude mice. Neither DP1 + ras nor E2F1 + ras is transforming in this assay (JOHNSON et al. 1994). Interestingly, activated ras + the DNA binding domain of E2F1 fused to the acidic transactivation domain of herpes simplex virus VP16 (E2F1-VP16) is transforming in the absence of DP1. This is consistent with the biological activity of E2F1 being repressed by its interaction with pocket proteins (FLEMINGTON et al. 1993; HELIN et al. 1993). However, it additionally suggests that the C-terminal region of E2F1 affects the requirement for DP1 in transformation, despite the fact that this region is not required for interaction with DP1 in vitro nor in vivo (BANDARA et al. 1993; HELIN et al. 1993; KREK et al. 1993). None of the combinations of E2F1, E2F1-VP16, DP1 and ras is as potent in transformation as E1a+ras.

In a third study XU et al. (1995) demonstrated that retroviral infection of NIH 3T3 cells with E2F1, -2, or -3 and subsequent drug selection generates polyclonal cultures that are able to grow in soft agar and have higher saturation densities than the control cells. Mutational analysis of E2F1-expressing cells shows that transformation requires the DNA binding domain and is enhanced by deletion of the pRB-binding domain, consistent with pRB being a negative reulator of E2F1. E2F4 also has oncogenic potential. GINSBERG et al. (1994) showed that infection of NIH 3T3 cells with a mutant E2F4 containing a four amino acid deletion in the pocket binding motif, but not wild-type E2F4, confers the ability to grow in soft agar. Likewise, another study provides evidence to suggest that E2F4, DP1, and activated ras cooperate in transformation of rat embryo fibroblasts (BEIJERSBERGEN et al. 1994). Taken together such results suggest that deregulated E2F expression or mutation of E2F so as to release it from negative regulation by pocket proteins is potentially oncogenic. There are as yet no reports of E2F family members being activated or overexpressed in human tumors, although SAITO et al. have shown the E2F1 gene to be amplified and overexpressed in the human erythroleukemia (HEL) cell line and translocated to abnormal chromosomal locations in several other established human leukemia cell lines (SAITO et al. 1995).

7 Conclusions and Future Questions

The functional studies in SAOS-2 cells have provided evidence consistent with the earlier biochemical data which suggested that E2F1 is regulated primarily by pRB, and E2F4 by p130. However, a number of questions still remain to be addressed. Firstly, more work is required to further characterize the cell cycle arrests imposed by pRB and p130. Do these arrests occur at the same, or different, points in the cell cycle? It is tempting to speculate that the pRB induced arrest is a G_1/S arrest and the p130 arrest is a G_0/G_1 arrest since p130 appears to be the major pocket protein present in E2F DNA-binding activity of quiescent fibroblasts (COBRINIK et al. 1993). Does the override of the p130 arrest by E2F4 require E2F4's ability to act as a transcription factor? Which putative E2F target genes, either alone or in combination, mimic the action of particular E2Fs and override a pocket protein induced arrest? Is there any specificity with regard to which putative E2F target genes override particular pocket protein induced arrests? Secondly, to what extent does the p107-induced arrest depend upon its observed in vivo association with E2F? Of note in this regard, evidence has recently been presented to suggest that p107 suppresses growth through both E2F-dependent and independent mechanisms (SMITH and NEVINS 1995; ZHU et al. 1995). Furthermore, as mentioned above, c-myc also associates with p107 in vivo and can override a p107-mediated growth arrest (BEIJERSBERGEN et al. 1994; GU et al. 1994). Thus, the precise molecular basis of the p107-mediated growth arrest is unclear at present. Equally obscure are the roles of the E2F/p107 complexes containing cyclin/cdks. Research in both areas should shed more light on the function of p107. Finally, it will be interesting to test the abilities of other cloned E2Fs to override such arrests. For example, the recently cloned E2F5 binds preferentially to p130 and therefore, as E2F4, may be expected to override a p130-induced arrest (HIJMANS et al. 1995). Likewise, E2F2 and E2F3 may be expected to override a pRB-induced arrest (IVEY-HOYLE et al. 1993; LEES et al. 1993).

The results demonstrating E2F1-stimulated override of cell cycle arrests induced by serum deprivation and TGFβ treatment provide evidence in support of the proposed role of E2F1 as a regulator of entry into S phase. Obviously there is a need to test the other E2Fs in such assays and determine whether there is any specificity with regard to a particular E2Fs ability to deregulate a particular cell cycle transition. The inducible cell lines described earlier provide useful model systems in which to further examine the molecular basis of E2F activity and also to illuminate the mechanisms of cell cycle control in general. For example, after overexpression of E2F1 in serum-deprived, quiescent fibroblasts has stimulated entry into S phase, what underlies the decision to undergo apoptosis as opposed to proliferation? Is p53 an upstream sensor and decision maker of the apoptotic process, or does it operate further downstream after the decision has been made? At what point relative to passage through S phase is the decision to die made?

It is important to consider the results demonstrating that E2Fs have oncogenic potential in the context of those demonstrating that deregulated E2F activity terminates in p53-dependent apoptosis. According to the model discussed above and illustrated in Fig. 2, deregulated E2F activity promotes S phase entry followed by apoptosis. This apoptotic pathway appears to be at least in part p53 dependent; although it is possible that other p53-independent pathways also exist (QIN et al. 1994). For deregulated E2F activity to give rise to transformation it is necessary to block, through secondary mutations, this apoptotic process. This model is therefore consistent with the idea that neoplasms arise as a result of a number of genetic changes. The apparent rarity of E2F mutations associated with human cancer may result from primary cells having efficient and multiple protective mechanisms against deregulated E2F activity.

Finally, studies on mice that have had one or more E2Fs functionally inactivated, and any cell lines that can be derived from such mice, will provide results complementary to those from overexpression studies. Such results are eagerly awaited.

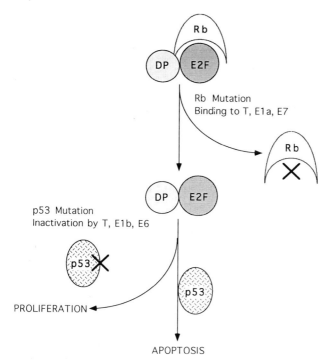

Fig. 2. Model for cooperation between inactivation of pRB and p53 in transformation. Functional inactivation of pRB, through mutation or binding to the viral oncoproteins, Ela, T, or E7, results in deregulated E2F activity and this stimulates entry into S phase. In the presence of wild-type p53 activity, this ultimately leads to apoptosis. However, functional inactivation of p53, through mutation or by the viral oncoproteins, E1b, T, or E6, inhibits the apoptotic pathway, thus promoting proliferation and transformation

Acknowledgments. We thank René Bernards, David Livingston, Ed Harlow, Claude Sardet, Gino Vairo, Bob Weinberg, and Liang Zhu for making data available prior to publication. We also thank members of the Kaelin, Livingston, Ewen, and DeCaprio laboratories for many stimulating thoughts and ideas.

References

Bandara LR, Buck VM, Zamaian M, Johnston LH, La Thangue NB (1993) Functional synergy between DP-1 and E2F-1 in the cell cycle-regulating transcription factor DRTF1/E2F. EMBO J 12: 4317–4324

Bartkova J, Lukas J, Muller H, Lutzhoft D, Strauss M, Bartek J (1994) Cyclin D1 protein expression and function in human breast cancer. Int J Cancer 57: 353–361

Beijhersbergen RL, Kerkhoven RM, Zhu L, Carlee L, Voorhoeve PM, Bernards R (1994a) E2F-4, a new member of the E2F gene family, has oncogenic activity and associates with p107 in vivo. Genes Dev 8: 2680–2690

Beijersbergen RL, Hijmans EM, Zhu L, Bernards R (1994b) Interaction of c-myc with the Rb related protein p107 results in inhibition of c-myc mediated transactivation. EMBO J 13: 4080–4086

Clarke AR, Maandag ER, van Roon M, van der Lugt NMT, van der Valk M, Hooper ML, Berns A, Riele H (1992) Requirement for a functional Rb-1 gene in murine development Nature 359: 328–330

Cobrinik D, Whyte P, Peeper DS, Jacks T, Weinberg RA (1993) Cell cycle-specific association of E2F with the p130 E1A-binding protein. Genes Dev 7: 2392–2404

Dalton S (1992) Cell cycle regulation of the human cdc2 gene. EMBO J 11: 1797–1804

Debbas M, White E (1993) Wild-type p53 mediates apoptosis by E1A, which is inhibited by E1B. Genes Dev 7: 546–554

Dynlacht BD, Flores O, Lees JA, Harlow E (1994) Differential regulation of E2F transactivation by cyclin/cdk complexes. Genes Dev 8: 1772–1786

Ewen ME (1994) The cell cycle and the retinoblastoma family. Cancer Metastaris Rev 13: 45–66

Ewen ME, Sluss HK, Whitehouse LL, Livingston DM (1993) TGFβ1 inhibition of cdk4 synthesis is linked to cell cycle arrest Cell. 74: 1009–1020

Ewen ME, Oliver C, Sluss H, Miller S, Peeper D (1995) p53-Dependent repression of CDK4 translation in TFG-β-induced G_1 cell-cycle arrest. Genes Dev 9: 204–217

Flemingron EK, Speck SH, Kaelin WG (1993) E2F-1 mediated transactivation is inhibited by complex formation with the retinoblastoma susceptibility gene product. Proc Natl Acad Sci USA 90: 6914–6918

Ginsberg D, Vairo G, Chittenden T, Xiao Z-X, Xu G, Wydner KL, DeCaprio JA, Lawrence JB, Livingston DM (1994) E2F-4 a new member of the E2F transcription factor family, interacts with p107 Genes. Dev 8: 2665–2679

Girling R, Bandara LR, Ormondroyd E, Lam EW-F, Kotecha S, Mohun T, La Thangue NB (1994) Molecular characterisation of Xenopus laevis DP proteins. Mol Biol Cell 5: 1081–1092

Goodrich DW, Lee W-H (1992) Abrogation by c-myc of G_1 phase arrest induced RB protein but not by p53. Nature 360: 177–179

Gu W, Bhatia K, Magrath IT, Dang CV, Dalla-Favera R (1994) Binding and suppression of the myc transcriptional activation domain by p107. Science 264: 251–254

Guan K-L, Jenkins CW, Li Y, Nichols MA, Wu X, O'Keefe CL, Matera AG, Xiong Y (1994) Growth suppression by p18, a p16 INK4/MTS1 and p14 INK4B/MTS1-related CDK6 inhibitor, correlates with wild-type pRb function. Genes Dev 8: 2939–2952

Hamel PA, Gill RM, Phillips RA, Gallie BL (1992) Transcriptional repression of the E2-containing promoters EllaE, c-myc, and RB1 by the product of the RB1 gene. Mol Cell Biol 12: 3431–3438

Hamel PA, Phillips RA, Muncaster M, Gallie BL (1993) Speculations on the role of RB1 in tissue specific differentiation, tumour initiation and tunour progression. FASEB J 7: 846–854

Hannon GJ, Beach D (1994) p15INK4B is a potential effector of TGF-β-induced cell cycle arrest. Nature 371: 257–261

Hannon GJ, Demetrick D, Beach D (1993) Isolation of the Rb-related p130 through its interactions with CDK2 and cyclins. Genes Dev 7: 2378–2391

Helin K, Lees JA, Vidal M, Dyson N, Harlow E, Fattaey A (1992) A cDNA encoding a pRB-binding protein with properties of the transcription factor E2F. Cell 70: 337–350

Helin K, Harlow E, Fattaey A (1993a) Inhibition of E2F-1 transactivation by direct binding of the retinoblastoma gene. Mol Cell Biol 13: 6501–6508

Helin K, Wu C-L, Fattaey AR, Lees JA, Dynlacht BD, Ngwu C, Harlow E (1993b) Heterodimerization of the transcription factors E2F-1 and DP-1 leads to cooperative transactivation. Genes Dev 7: 1850–1861

Herber B, Truss M, Beato M, Muller R (1994) Inducible regulatory elements in the cyclin D1 promoter. Oncogene 9: 1295–1304

Hiebert SW, Chellappan SP, Horowitz JM, Nevins JR (1992) The interaction of RB with E2F coincides with an inhibition of the transcriptional activity of E2F. Genes Dev 6: 177–185

Hijmans EM, Voorhoeue PM, Beijersbergen RL, van't Veer LJ, Bernards R (1995) E2F-5, a new E2F family member that interacts with p130 in vivo. Mol Cell Biol 15: 3082–3089

Hinds PW, Mittnacht S, Dulic V, Arnold A, Reed SI, Weinberg RA (1992) Regulation of retinoblastoma protein functions by ectopic expression of human cyclins. Cell 70: 993–1006

Howe JA, Mymryk JS, Egan C, Branton PE, Bayley ST (1990) Retinoblastoma growth suppressor and a 300-kDa protein appear to regulate cellular DNA synthesis. Proc Natl Acad Sci USA 87: 5883–5887

Howes KA, Ransom N, Papermaster DS, Lasudry JGH, Albert DM, Windle JJ (1994) Apoptosis or retinoblastoma: alternative fates of photoreceptors expressing the HPV-16 E7 gene in the presence or absence of p53. Genes Dev 8: 1300–1310

Hsiao K-M, McMahon SL, Farnham P (1994) Multiple DNA elements are required for the growth regulation of the mouse E2F1 promoter. Genes Dev 8: 1526–1537

Huang H-JS, Yee J-K, Shew J-Y, Chen P-L, Bookstein R, Friedmann T, Lee EY-HP, Lee W-H (1988) Suppression of the neoplastic phenotype by replacement of the RB gene in human cancer cells. Science 242: 1563–1566

Huber HE, Edwards G, Goodhart PJ, Patrick DR, Huang PS, Ivey-Hoble M, Barnett SF, Oliff A, Heimbrook DC (1993) Transcription factor E2F binds DNA as a heterodimer. Proc Natl Acad Sci USA 90: 3525–3529

Ivey-Hoyle M, Conroy R, Huber HE, Goodhart PJ, Oliff A, Heimbrook DC (1993) Cloning and characterization of E2F-2, a novel protein with the biochemical properties of transcription factor E2F. Mol Cell Biol 13: 7802–7812

Jacks T, Fazeli A, Schmitt EM, Bronson RT, Goodell MA, Weinberg RA (1992) Effects of an Rb mutation in the mouse. Nature 359: 295–300

Johnson D, Cress W, Jakoi L, Nevins J (1994a) Oncogenic capacity of the E2F1 gene. Proc Natl Acad Sci USA 91: 12823–12827

Johnson DG, Ohtani K, Nevins JR (1994b) Autoregulatory control of E2F1 expression in response to positive and negative regulators of cell cycle progression. Genes Dev 8: 1514–1525

Johnson DG, Schwarz JK, Cress WD, Nevins JR (1993) Expression of transcription factor E2F1 induces quiescent cells to enter S phase. Nature 365: 349–352

Joliff K, Li Y, Johnson LF (1991) Multiple protein DNA interactions in the TATA less mouse thymidylate synthase promoter. Nucleic Acids Res 19: 2267–2274

Kaelin WG Jr, Krek W, Sellers WR, DeCaprio JA, Ajchenbaum F, Fuchs CS, Chittenden T, Li Y, Farnham PJ, Blanar MA, Livingston DM, Flemington EK (1992) Expression cloning of a cDNA encoding a retinoblastoma-binding protein with E2F-like properties. Cell 70: 351–364

Kovesdi I, Reichel R, Nevins JR (1986) Identification of a cellular transcription factor involved in E1A trans-activation. Cell 45: 219–228

Krek W, Livingston DM, Shirodkar S (1993) Binding to DNA and retinoblastoma gene promoted by complex formation of different E2F family members. Science 262: 1557–1560

Krek W, Ewen M, Shirodkar S, Arany Z, Kaelin WG Jr, Livingston DM (1994) Negative regulation of the growth-promoting transcription factor E2F-1 by a stably bound cyclin A-dependent protein kinase. Cell 78: 1–20

La Thangue NB, Rigby PWJ (1987) An adenovirus E1A-like transcription factor is regulated during the differentiation of murine embryonal carcinoma stem cells. Cell 49: 507–513

Laiho M, DeCaprio JA, Ludlow JW, Livingston DM, Massagué J (1990) Growth inhibition by TGF-β linked to suppression of retinoblastoma protein phosphorylation. Cell 62: 175–185

Lam EW-F, Watson RJ (1993) An E2F-binding site mediates cell-cycle regulated repression of mouse B-Myb transcription. EMBO 12: 2705–2713

Lee EY-HP, Chang C-Y, Hu N, Wang Y-CJ, Lai C-C, Herrup K, Lee W-H, Bradley A (1992) Mice deficient for Rb are nonviable and show defects in neurogenesis and haematopoiesis. Nature 359: 288–294

Lee EY-HP, Hu N, Yuan S-SF, Cox LA, Bradley A, Lee W-H, Herrup K (1994) Dual roles of the retinoblastoma protein in cell cycle regulation and neuron differentiation. Genes Dev 8: 2008–2021

Lees JA, Saito M, Vidal M, Valentine M, Look T, Harlow E, Dyson N, Helin K (1993) The retinoblastoma protein binds to a family of E2F transcription factors. Mol Cell Biol 13: 7813–7825

Li Y, Slansky JE, Myers DJ, Drinkwater NR, Kaelin WG, Farnham PJ (1993) Cloning chromosomal location and characterization of mouse E2F1. Mol Cell Biol 14: 1861–1869

Lowe SW, Ruley HE (1993) Stabilisation of the tumour suppressor p53 is induced by adenovirus E1a and accompanies apoptosis. Genes Dev 7: 535–545

Lukas J, Muller H, Bartkova J, Spitkovsky D, Kjerulff AA, Jansen-Durr P, Strauss M, Bartek J (1994) DNA tumor virus oncoproteins and retinblastoma gene mutations share the ability to relieve the cell's requirement for cyclin D1 function in G1. J Cell Biol 125: 625–638

Mittnacht S, Weinberg RA (1991) G₁/S phosphorylation of the retinoblastoma protein is associated with an altered affinity for the nuclear compartment. Cell 65: 381–393

Moran B, Zerler B (1988) Interactions between cell growth-regulating domains in the products of the adenovirus E1A oncogene. Mol Cell Biol 8: 1756–1764

Moran E (1993) DNA tumor virus transforming proteins and the cell cycle. Curr Opin Genet Dev 3: 63–70

Morgenbesser SD, Williams BO, Jacks T, DePinho RA (1994) p53 dependent apoptosis produced by Rb deficiency in the developing mouse lens. Nature 371: 72–74

Muller H, Lukas J, Schneider A, Warthoe P, Bartek J, Eilers M, Strauss M (1994) Cyclin D1 expression is regulated by the retinoblastoma protein. PNAS 91: 2945–2949

Mymryk JS, Shire K, Bayley ST (1994) Induction of apoptosis by adenovirus type E1a in rat cells requires a proliferation block. Oncogene 9: 1188–1193

Neuman E, Flemington EK, Sellers WR, Kaelin WG (1994) Transcription of the E2F1 gene is rendered cell-cycle dependent by E2F DNA-binding sites within its promoter. Mol Cell Biol 14: 6607–6615

Nevins JR (1992) E2F: a link between the Rb tumor suppressor protein and viral oncoproteins. Science 258: 424–429

Oswald F, Lovec H, Moroy T, Lipp M (1994)E2F dependent regulation of human myc: transactivation by cyclins D1 and A overrides tumour supressor protein functions. Oncogene 9: 2029–2036

Pan H, Griep AE (1994) Altered cell cycle regulation in the lens HPV-16 or E7 transgenic mice: implications for tumour suppressor gene function in development. Genes Dev 8: 1285–1299

Pearson BE, Nasheuer HP, Wang TSF (1991) Human DNA polymerase α gene, sequences controlling expression in cycling and serum stimulated cells. Mol Cell Biol 11: 2081–2095

Peeper DS, Keblusek P, Helin K, Toebes M, Van der Eb AJ, Zantema A (1995) Phosphorylation of a specific site in E2F1 affects its electrophoretic mobility and promotes pRb binding in vitro. Oncogene 10: 39–48

Philipp A, Schneider A, Vasrik I, Finke K, Xiong Y, Beach D, Alitalo K, Eilers M (1994) Repression of cyclin D1: a novel function of MYC. Mol Cell Biol 14: 4032–4043

polyaak K, Lee M-H, Erdjument-Bromage H, Koff A, Roberts J, Tempst P, Massague J (1994) Cloning of p27Kip1, a cyclin-dependent kinase inhibitor and a potential mediator of extracellular anti-mitogenic signals. Cell 78: 59–66

Qian Y, Luckey C, Horton L, Esser M, Templeton DJ (1992) Biological function of the retinoblastoma protein requires distinct domains for hyperphosphorylation and transcription factor binding. Mol Cell Biol 12: 5363–5372

Qin X-Q, Chittenden T, Livingston DM, Kaelin WG (1992) Identification of a growth suppression domain within the retinoblastoma gene product. Genes Dev 6: 953–964

Qin X-Q, Livingston DM, Kaelin WG, Adams P (1994) Deregulated E2F1 expression leads to S-phase entry and p53-mediated apoptosis. Proc Natl Acad Sci USA 91: 10918–10922

Qin X-Q, Livingston DM, Ewen M, Sellers WR, Arany Z, Kaelin WG (1995) The transcription factor E2F1 is a downstream target of RB action. Mol Biol Cell 15: 742–755

Roussel MF, Davis JN, Cleveland JL, Ghysdael J, Hiebert SW (1994) Dual control of myc expression through a single DNA binding site targeted by ets family proteins and E2F-1. Oncogene 9: 405–415

Saenz Robles MT, Symonds H, Chen J, Van Dyke T (1994) Induction versus progression of brain tumour development: differential functions for the Rb and p53 targetting domains of the SV40 T antigen. Mol Cell Biol 14: 2686–2698

Saito M, Helin K, Valentine MB, Griffith BB, Willman CL, Harlow E, Look AT (1995) Amplification of the E2F1 transcription factor gene in the HEL Erythroleukemia cell line. Genomics 25: 130–138

Sardet C, Vidal M, Cobrinik D, Geng Y, Onufryk C, Chen A, Weinberg RA (1995) E2F4 and E2F5, two novel members of the E2F family, are expressed in the early phases of the cell cycle. PNAS 92: 2403–2407

Schwarz JK, Bassing CH, Kovesdi I, Datto MB, Blazing M, George S, Wang X-F, Nevins JR (1995) Expression of the E2F1 transcription factor overcomes TGFb mediated growth suppression. PNAS 92: 483–487

Sellers WR, Rodgers J, Kaelin WG, Jr (1995) A potent transrepression domain in the retinoblastoma protein induces cell-cycle arrest when bound to E2F sites. Proc Natl Acad Sci USA (in press)

Shan B, Lee W-H (1994) Deregulated expression of E2F-1 induces S-phase entry and leads to apoptosis. Mol Cell Biol 14: 8166–8173

Shan B, Zhu X, Chen P-L, Durfee T, Yang Y, Sharp D, Lee W-H (1992) Molecular cloning of cellular genes encoding retinoblastoma-associated proteins: identification of a gene with properties of the transcription factor E2F. Mol Cell Biol 12: 5620–5631

Singh P, Wong S, Hong W (1994) Overexpression of E2F-1 in rat embryo fibroblasts leads to neoplastic transformation. EMBO J 13: 3329–3338

Smith EJ, Nevins JR (1995) The Rb related p107 protein can suppress E2F function independently of binding to Cyclin A/cdk2. Mol Cell Biol 15: 338–344

Stein R, Corrigan M, Yaciuk P, Whelan J, Moran E (1990) Analysis of E1A-mediated growth regulation functions: binding of the 300-kilodalton cellular product correlates with E1A enhancer repression function and DNA synthesis-inducing activity. J Virol 64: 4421–4427

Symonds H, Krall L, Remington L, Saenz-Robles M, Lowe S, Jacks T, Van Dyke T (1994) p53 dependent apoptosis suppresses tumour growth and progression in vivo. Cell 78: 703–711

Vairo G, Livingston DM, Ginsberg D (1995) Functional Interaction between E2F4 and p130: evidence for distinct mechanisms underlying growth suppression by different Rb family members. Genes Dev 9: 869–881

Weintraub SJ, Prater CA, Dean DC (1992) Retinoblastoma protein switches the E2F site from positive to negative element. Nature 358: 259–261

White E, Cipriani R, Sabbatini P, Denton A (1991) Adenovirus E1B 19-kilodalton protein overcomes the cytotoxicity of E1A proteins. J Virol 65: 2968–2978

Wu C-L, Zukerberg LR, Ngwu C, Harlow E, Lees JA (1995) In vivo association of E2F and Dp family proteins. Mol Cell Biol 15: 2536–2546

Wu X, Levine AJ (1994) p53 and E2F1 Cooperate to Mediate Apoptosis. Proc Natl Acad Sci USA 91: 3602–3606

Xu G, Livingston D, Krek W (1995) Multiple members of the E2F transcription factor family are the products of oncogenes. Proc Natl Acad Sci USA 92: 1357–1361

Xu M, Sheppard KA, Peng C-Y, Yee AS, Piwnica-Worms H (1994) Cyclin A/cdk2 binds directly to E2F1 and inhibits the DNA-binding activity of E2F1/DP1 by phosphorylation. Mol Cell Biol 14: 8420–8431

Yamamoto M, Yoshida M, Ono K, Fujita T, Ohtani-Fujita N, Sakai T, Nikaido T (1994) Effect of tumor suppressors on cell cycle-regulatory genes. Exp Cell Res 210: 94–101

Zamanian M, La Thangue NB (1992) Adenovirus E1A prevents the retinoblastoma gene product from repressing the activity of a cellular transcription factor. EMBO J 11: 2603–2610

Zhu L, van der Heuvel S, Helin K, Fattaey A, Ewen M, Livingston D, Dyson N, Harlow E (1993) Inhibition of cell proliferation by p107, a relative of the retinoblastoma protein. Genes Dev 7: 1111–1125

Zhu L, Enders G, Lees JA, Beijersbergen RL, Bernards R, Harlow E (1995) The pRB related protein p107 contains two growth suppression domains: independent interactions with E2F and cyclin/cdk complexes. EMBO J 14: 1904–1913

Start-Specific Transcription in Yeast

L. Breeden

1 Introduction

At the G_1 to S transition of the budding yeast cell cycle there is burst of transcription of at least 30 different genes. This may in part be due to the fact that *Saccharomyes cerevisiae* often exists as a colonial micro-organism which spends most of its time in stationary phase (G_0). When cells have the opportunity to enter the mitotic cell cycle, there is a selective advantage for cells that can start dividing rapidly and efficiently. Thus it is not surprising that they resynthesize enzymes critical for high-fidelity replication of DNA and replace other components that may

Fred Hutchinson Cancer Research Center, 1124 Columbia St., Seattle, WA 98104, USA not have survived extended G_0 arrest. Proteins that are responsible for starting the cell cycle, particularly those that would disrupt the cycle if they were produced at other stages of the cell cycle, are transcribed specifically at the G_1/S transition. Most of the G_1 cyclins are expressed specifically at this time, and their activity determines the timing of the G_1/S transition (RICHARDSON et al. 1989; NASH et al. 1988; CROSS 1988).

The regulation of G_1/S-specific gene expression is exerted primarily at the level of transcription initiation. The promoter elements that are known to be specifically active at the G_1/S border are called SCBs and MCBs (see Sect. 2, 3). Tandem oligomers of SCB and MCB elements are sufficient to confer G_1/S-specific transcription to heterologous genes (BREEDEN and NASMYTH 1987a; MCINTOSH et al. 1991; GORDON and CAMPBELL 1991). These are the binding sites for the Swi4/Swi6 and Mbp1/Swi6 complexes (ANDREWS and HERSKOWITZ 1989b; NASMYTH and DIRICK 1991; LOWNDES et al. 1992b; DIRICK et al. 1992). Swi4 and Mbp1 are highly related to each other, especially in their N-terminal DNA binding domains (ANDREWS and HERSKOWITZ 1989b; KOCH et al. 1993). Swi6, Swi4 and Mbp1 also contain five tandem copies of a 33 amino acid repeat which has been referred to as a Swi6/Cdc10, TPLH, or ankyrin repeat (BREEDEN and NASMYTH 1987b; BORK 1993).

Although many fewer G_1/S-specific transcripts have been identified in *S. pombe*, their mechanism of regulation has been conserved. MCB elements are responsible for the periodic transcription of *cdc22* in *S. pombe* (LOWNDES et al. 1992a), and these sites are bound by proteins structurally related to the *S. cerevisiae* gene products (see Fig. 1). Cdc 10 (LOWNDES et al. 1992a) associates with either Res1 (TANAKA et al. 1992; CALIGIURI and BEACH 1993) or Res2 (ZHU et al. 1994; MIYAMOTO et al. 1994). The Res proteins confer the DNA binding specificity to the complex through N-terminal domains which are homologous to those of Swi4 and Mbp1. *Kluveromyces lactis* homologs for Mbp1 and Swi6 have been identified (KOCH et al. 1993), and *Absidia zychae*, another pathogenic fungus, also encodes a protein related to Mbp1 (L. Breeden, unpublished).

If we consider other proteins that share only the DNA binding motif found in the Swi4/6 family, there are five additional members of this group (see Fig. 2): (a) Phd1 is a protein whose overexpression promotes precocious pseudohyphal growth of *S. cerevisiae* (GIMENO and FINK 1994). This is a highly branched, invasive form of growth that is normally induced by nitrogen starvation. (b) The Efg1 protein of *Candida albicans* (EMBL X71621) contains a DNA binding domain very closely related to that of Phd1 and could play an analogous role in this invasive human pathogen. (c) Sok2 was isolated as a gene whose overexpression partially suppresses a protein kinase A defect (WARD and GARRETT 1994), which has been sequenced recently (Ward and Garrett, unpublished). (d) 647 is a putative open reading frame of unknown function in *S. cerevisiae* (L. Breeden unpublished). (e) The StuA protein of *Aspergillus nidulans* was the first of this group to be identified. It is required for the normal development of conidiophores (MILLER et al. 1991, 1992). These five genes are almost certainly transcription factors, but their gene targets and regulation are not understood. It will be interesting to determine whether there is any cross-talk between these transcription factors and members of the Swi4/6 family of proteins.

So far all the members of this extended family of proteins are found in fungi. It is not clear whether they are unique to fungi or have merely eluded detection in higher cells. It is possible that the closest functional equivalent of the Swi4/6 family in higher cells is the E2F/DP1 family of transcription complexes (for review see LA THANGUE 1994, and the other chapters of this volume). There is very little homology at the sequence level (LA THANGUE and TAYLOR 1993), but many functional parallels can be found between these protein families. Remarkably, the E2F/DP1 binding site closely resembles the SCB and MCB elements of *S. cerevisiae*, and these elements all activate transcription at the beginning of S phase (MEAN et al. 1992; MERRILL et al. 1992). Rb binds and represses the transcriptional activity of the E2F complexes (HIEBERT et al. 1992), and Swi6 appears to be the repressor in the Swi4/6 complex (BREEDEN and MIKESELL 1994). Both the Rb and Swi6 proteins are phosphorylated in a cell cycle dependent manner (DECAPRIO et al. 1992; Sidorova and Breeden to be published). There is a family of E2F-related factors (LEES et al. 1993; IVEY-HOYLE et al. 1993), and E2F1, like Swi4, is also cell cycle regulated at the transcription level (SHAN et al. 1992).

This review focuses on the Swi4/6 family, which includes Swi4, Swi6, Mbp1, Cdc10, Res1, and Res2. These transcription factors are responsible for most of the transcription that occurs at the Start of the cell cycle in *S. cerevisiae* and *Schizosaccharomyces pombe*. The first sections describe how they were identified, and what is known about their functional domains. Later sections focus on how they are regulated and explore the significance of this pathway on growth regulation.

2 Identification of the SCB Binding Complex

The Swi4/6 binding complex was the first cell cycle regulated transcription complex to be elucidated in model studies of the *HO* promoter of *S. cerevisiae*. *HO* encodes the double strand endonuclease that initiates mating type switching (KOSTRIKEN et al. 1983). Early studies of the pattern of mating type switching suggested that switching is a cell cycle regulated process (STRATHERN and HERSKOWITZ 1979) and this was shown to be due to the regulated transcription of *HO* (NASMYTH 1983). Transcripts of the *HO* gene accumulate only during the late G_1– early S phase of the cell cycle, and promoter fusions to the *MATα 1* gene were used to show that the *HO* promoter can confer this regulation on a completely heterologous transcript (NASMYTH 1985a). These results indicate that initiation of transcription is cell cycle regulated. Deletion analysis of the *HO* promoter showed that the cell cycle regulatory element is redundant, and indeed there are eight copies of the sequence CACGAAA in a 700-bp region of the *HO* promoter (NASMYTH 1985b). This sequence, now called an SCB, or Swi4/6-dependent cell cycle box, was shown to be an upstream activation sequence (UAS) sufficient to confer G_1/S-specific transcription to heterologous genes (BREEDEN and NASMYTH 1987a; ANDREWS and HERSKOWITZ 1989b).

To identify the *trans*-acting factors responsible for the UAS activity of SCB elements, a mutant search was carried out with a yeast strain carrying an *HO::lacZ* fusion (BREEDEN and NASMYTH 1987a) and a colorimetric filter assays for detecting LacZ+ colonies (BREEDEN and NASMYTH 1985). Mutants that were LacZ– because they could not transcribe *lacZ* from the *HO* promoter were identified and put into complementation groups. They were called *swi* mutants because a subset of them were allelic with mutants previously isolated for the inability to switch mating types (HABER and GARVIK 1977; STERN et al. 1984). Representatives from each *swi* complementation group were tested for the ability to activate transcription from tandem SCB elements, and *SWI4* and *SWI6* were identified as genes that are specifically required for SCB activation (BREEDEN and NASMYTH 1987a). The cloning of *SWI6* yielded two different complementing clones encoding the gene for *SWI6* and a suppressor identified as *SWI4* (BREEDEN and NASMYTH 1987b). The discovery that overproduction of Swi4 can suppress the requirement for Swi6 was the first evidence that Swi4 is the primary activator, and that Swi6 plays an accessory role in SCB activation.

The sequence of *SWI6* revealed homology to the $cdc10^+$ gene, which is its counterpart in *S. pombe* (BREEDEN and NASMYTH 1987b). $cdc10^-$ mutants were isolated as temperature sensitive (ts) lethal mutants that arrested cells in G_1 (NURSE et al. 1976). When arrested, $cdc10^-$ cells display properties characteristic of G_1 cells such as continued growth and mating capacity, yet they are unable to start the mitotic cell cycle (NURSE and BISSETT 1981). This phenotype is similar to that observed during the G_1 arrest conferred by *cdc28* in *S. cerevisiae* (REID and HARTWELL 1977), and $cdc2^-$ in *S. pombe* (NURSE and BISSETT 1981). The homology between *SWI6* and $cdc10^+$ extends over the latter two-thirds of their coding sequences over which they are about 30% identical (BREEDEN and NASMYTH 1987b), and several blocks of homology are shown in Figs.2–4. Despite this low level of homology Swi6 and Cdc10 play related roles in these two highly divergent yeast. One important difference is that the *swi6* deletion is not lethal (BREEDEN and NASMYTH 1987b), whereas $cdc10^+$ is an essential gene (NURSE et al. 1976). Both gene products activate transcription of essential genes. The difference seems to lie in the existence of alternative pathways for expression of these essential genes in *S. cerevisiae*.

Deletions of *swi4* are not lethal in the original strain background that was tested, but the *swi4 swi6* double mutant is lethal, and both single mutants grow slowly and are abnormally large (BREEDEN and NASMYTH 1987b). Since *HO* is a completely nonessential gene, these pleiotropic effects indicate that Swi4 and Swi6 are required for the transcription of other genes. *swi4* mutants were later found to be lethal in some strain backgrounds, and the mutants cause a predominantly G_1 arrest (OGAS et al. 1991). Searches for high copy suppressors of this lethality (OGAS et al. 1991), and the lethality of the *swi4ts swi6* double mutant (NASMYTH and DIRICK 1991) led to the discovery that ectopic expression of G_1 cyclins

[Cln1, Cln2, and Pcl1 (Hcs26)] can rescue these mutants. SCB elements were found in the *CLN2* and *PCL1* promoters, and Swi4 and Swi6 were shown to bind specifically to these elements. Subsequent to these studies a fourth cyclin Pcl2 (OrfD) was identified by its homology to Pcl1. All four of these cyclin transcripts peak at the G_1/S boundary (WITTENBERG et al. 1990; TYERS et al. 1993; MEASDAY et al. 1994).

These G_1 cyclin mRNAs fluctuate in the cell cycle, and peak at the same time that *HO* mRNA peaks. G_1 cyclin transcript levels are lower in *swi4* and *swi6* mutants, but they are not eliminated as is the case with *HO* (BREEDEN and NASMYTH 1987a). This suggests that there are parallel paths for activating the cyclin promoters. The *CLN1* promoter contains no good matches to the SCB consensus (OGAS et al. 1991), but it is clearly regulated in a similar manner (WITTENBERG et al. 1990). Deletion analysis indicated that another DNA sequence, containing three MCB-like elements, is responsible for most of the cell cycle regulation conferred upon this promoter. Surprisingly, the predominant binding activity observed on that site is a Swi4/Swi6 complex, not the expected Mbp1/Swi6 complex (Partridge and Breeden, to be published).

G_1 cyclins are required for the G_1/S transition of *S. cerevisiae* (RICHARDSON et al. 1989; ESPINOZA et al. 1994). The finding that Swi4/Swi6 complexes are involved in G_1 cyclin transcription means that understanding the regulation of Swi4/Swi6 activity is likely to provide new insight into the mechanism of Start, which is the critical control point in the *S. cerevisiae* cell cycle. So far there is no evidence of an SCB activation pathway in *S. pombe*, but the MCB activation pathway is conserved in fission yeast and may play an analogous role.

3 Identification of the MCB Binding Complex

Another large group of genes involved in DNA metabolism are also transcribed predominantly at the G_1/S boundary. The promoters of *TMP1*, *CDC8*, and *CDC9* were the first of this group to be characterized (WHITE et al. 1987). These promoters all contain at least one and often multiple MluI sites (ORD et al. 1988). ACGCGT (MluI) sequences are not common sequences in the AT-rich, intergenic DNA of *S. cerevisiae*, and further mutation analysis showed that they define a second regulatory element (McINTOSH et al. 1991; GORDON and CAMPBELL 1991). These elements are called MluI cell cycle boxes, or MCBs, and they are sufficient to confer cell cycle regulated transcription to heterologous genes (McINTOSH et al. 1991; LOWNDES et al. 1991). MCB elements were also found upstream from the *cdc22+* gene of *S. pombe*, which is a cell cycle regulated gene whose mRNA peaks early in the cell cycle (KELLY et al. 1993). *cdc22+* transcription is dependent upon Cdc10 function, and Cdc10 binds to these sites in *S. pombe* (LOWNDES et al. 1992a).

Cdc10 has two known binding partners in *S. pombe*: Res1 and Res2. Semi-dominant alleles of *res1* were isolated as suppressors of a *cdc10-ts* mutant (called *sct1*) (MARKS et al. 1993; CALIGIURI and BEACH 1993). *res1+* was also isolated as a high-copy suppressor of the growth arrest and meiosis caused by *pat1ts* muta-tions (MIYAMOTO et al. 1994). Res1 resembles Swi4 both in structure and function. Overproduction of *Res1+* also suppresses temperature sensitive mutants of *cdc10−* and *res1−* mutants are lethal in some strain background (CALIGIURI and BEACH 1993; MIYAMOTO et al. 1994). Res1 complexes with Cdc10 and binds to the MCB elements in the *cdc22+* promoter (CALIGIURI and BEACH 1993; AYTE et al. 1995), but may bind to a different sequence in the *cdt1* promoter (HOFMANN and BEACH 1994). *cdc18* is another periodically expressed gene that is dependent upon Cdc10 and its promoter contains many MCB-like sequences (KELLY et al. 1993).

Res2 is the most recently identified binding partner of Cdc10. As with Res1, Res2 can suppress *cdc10ts* mutations when it is overproduced, but *res 2+* was isolated as a high-copy suppressor of *res1−* (MIYAMOTO et al. 1994), and as an MCB binding partner of Cdc10 (called *pct1+*) (ZHU et al. 1994). *res2+* is not an essential gene, but *res2−* is lethal in combination with either *cdc10−* or *res1−* mutations. Thus, Res2 is functionally redundant with Res1 and is dispensable during mitotic growth. Res1 and Res2 both have roles during meiosis, because both mutants cause defects in meiosis and spore formation (MIYAMOTO et al. 1994; ZHU et al. 1994). Res2 transcription is induced during conjugation (MIYAMOTO et al. 1994), which immediately precedes meiosis in *S. pombe*. The *res1−* defect in meiosis can be suppressed by Res2 overproduction or by the normal induction of Res2 upon conjugation (MIYAMOTO et al. 1994; ZHU et al. 1994). Thus it is believed that Res1 plays the predominant role during the mitotic cell cycle, and Res2 predomi-nates during meiosis.

The discovery of the role of Cdc10 in MCB activation led immediately to the finding that Swi6 binds to MCB sites in *S. cerevisiae* (LOWNDES et al. 1992b; DIRICK et al. 1992). The binding partner of Swi6 at MCB sites was identified by its homology to Swi4 (KOCH et al. 1993). MCB binding protein or Mbp1 is a nonessen-tial gene, but the *mbp1* deletion is lethal in combination with *swi4* mutants (KOCH et al. 1993). This may reflect some redundancy between these MCB and SCB binding proteins in vivo.

Sequence analysis of many genes including the genes encoding cyto-chrome *c*, alcohol dehydrogenase, and histones has shown that *S. cerevisiae* genes are as different from the corresponding genes of *S. pombe* as they are from their human homologs (SIPICZKI 1989). This indicates that these two yeasts have undergone extensive evolutionary divergence. However, the MCB promoter element, the proteins that bind to it, and its time of activation within the cell cycle have all been conserved in budding and fission yeast. This is clearly a conserved pathway of transcriptional regulation and is not necessarily restricted to fungi. There has been one report that Cdc10 antibodies cross-react with a single human protein of similar molecular weight (SIMANIS and NURSE 1989), and the growing list of fungal homologs should make degenerate PCR an effective way to search for these genes in higher cells.

4 Anatomy of the Transcription Complexes

The basic anatomy of the Swi4/6 family of transcription factors is similar, as far as we know. They are at least heterodimeric complexes, with a very high apparent molecular weight on mobility shift gels. In this review they are referred to by their known and assayable components (e.g., Swi4/Swi6, Res1/Cdc10), with the recognition that they may contain other proteins. They are also referred to in the context of the specific binding site upon which they were assayed. This is necessary because we do not know how many different complexes there are, or what determines their specific binding site preferences (see Sect. 5). For example, the two known components that bind the SCB elements in the *HO* promoter are Swi4 and Swi6, and this complex has been referred to as SCB binding factor. However, Swi4 and Swi6 bind to MCB-like sites in the *CLN1* promoter (Partridge and Breeden, to be published), and could as easily also be called MCB binding factor.

4.1 DNA Binding Domain

In *S. cerevisiae* cells containing a normal level of Swi4 protein, Swi6 is absolutely required for *HO* transcription (BREEDEN and NASMYTH 1987a) and for formation of specific DNA protein complexes in mobility shift assays (ANDREWS and HERSKOWITZ 1989a). When Swi4 is overproduced, *HO* transcription can occur in the absence of Swi6 (BREEDEN and NASMYTH 1987b). This indicates that either Swi4 or an unidentified component of the Swi4/Swi6 complex must contain the DNA binding domain. In mobility shift experiments with crude extracts of *swi6* cells overproducing Swi4, Swi4 complexes on SCB elements can be observed, but they are heterogeneous in size and probably represent binding of proteolytic fragments of Swi4 to SCB sites (SIDOROVA and BREEDEN 1993). Mobility shift experiments using SCB elements from the *CLN2* promoter, and in vitro translated fragments of the Swi4 protein have shown that amino acid residues 36–155 contain a DNA binding activity specific for SCB and MCB elements (PRIMIG et al. 1992). However, there is no evidence that in vitro translated full-length Swi4 can form a discrete complex on DNA. Only a disperse array of low molecular weight complexes can be observed. These data are most consistent with the view that Swi4 contains the DNA binding domain at its amino terminus, but this region is inaccessible in the full-length Swi4, or it does not have sufficient affinity to bind SCBs until Swi6 is associated. Purified Swi6 shows no DNA binding specificity (SIDOROVA and BREEDEN 1993; KOCH et al. 1993), but it could contribute to the binding affinity of the complex.

The DNA binding domain of Swi4 is highly related to the amino-terminal domains of Mbp1 and Res1 (see Fig. 1), which have also been shown to bind DNA. Mbp1 binds DNA through its first 124 amino acid residues (DIRICK et al. 1992). The N-terminal fragment of Res1 is also sufficient for DNA binding in vitro

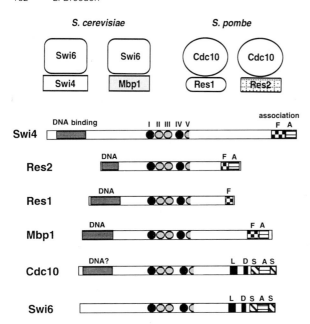

Fig. 1. Similarities within the Swi4/6 family of transcription factor. *Above*, the known associations between family members from *S. cerevisiae* and *S. pombe*; *below*, regions of homology between these family members. *Shaded boxes*, DNA binding domain homology (see also Fig. 2); *elipses*, consensus Swi6/Cdc10 repeats (*black*) and the degenerate repeats (*gray*) that are aligned in Fig. 3. The carboxy-terminal homologies (*F, A, L, D, S,* and *A*) are shared among family members as depicted in the drawing. These homologies are aligned in Fig. 4 and are within the region required for association

and in vivo (CALIGIURI and BEACH 1993; Ayte et al., to be published). A semi-dominant allele of *res1*, which was isolated as a suppressor of *cdc10-129*, has been sequenced and shown to carry a mutation in the DNA binding domain that changes the glutamic acid at position 56 to lysine (E56K; CALIGIURI and BEACH 1993). Since the E56K mutation suppresses *cdc10* mutations under conditions that the wild-type sequence does not suppress, it was suggested that lysine 56 increases the affinity for DNA in vivo. This has not yet been confirmed by in vitro studies, and it is therefore possible that E56K affects a different function of this domain. Although the original sequence alignments did not indicate this, the position analogous to E56 in the Swi4/6 family of proteins is always acidic (E or D), and the mutation converts it to a basic residue, which is what is found at the equivalent position in Phd1, Efg1, StuA, and Sok2 (see asterisk in Fig. 2). The only exceptions are ORF 647 and Res2, which do not contain any homology to this section of the proteins. Since the E to K change in Res1 allows it to bind and activate in the absence of Cdc10, it will be interesting to determine whether this mutation simply improves its DNA binding affinity, or whether it affects the regulation of Res1 activity. Res2 has no E56 region and therefore it cannot be essential for DNA binding.

One striking difference between the *S. pombe* and *S. cerevisiae* transcription complexes is that the N-terminus of Cdc10, unlike that of Swi6, also contains considerable homology to the Swi4 DNA binding domain (see Fig. 2; PRIMIG et al. 1992). In vitro translated Cdc10 does not bind DNA, as assayed on mobility shift gels (ZHU et al. 1994), but it is possible that there are some sites or conditions under which it can bind DNA. For example, one of Cdc10's binding partners, Res2,

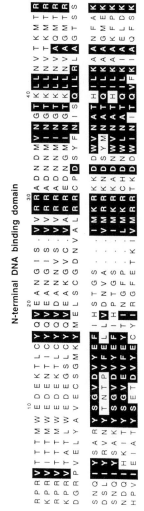

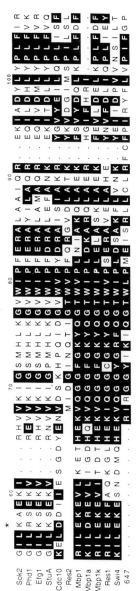

Fig. 2. Twelve fungal proteins share homology with the DNA binding domain first identified in Swi4. *Boxes,* identities and similarities (I, V, L) (F, Y) (E, D) (K, R) to the Swi4 DNA binding domain. *Asterisk,* the E56 position of Res1 which when changed to lysine (K) results in a semidominant, Cdc10-independent phenotype (CALIGIURI and BEACH 1993). Swi6k and Mbp1k denote the putative Swi6 and Mbp1 homologs from *K. lactis.* Mbp1a is a relative of Mbp1 from *A. zychae*

has a smaller region of homology to the DNA binding domain of its relatives. The Res2 protein lacks the first 70 residues of homology, including E56. Its homology begins just before the highly conserved GXWXP motif (see Fig. 2). This could indicate that the critical residues for DNA binding include only the GXWXP motif and beyond. However, overproduction of Res2 suppresses *cdc10* null mutants very inefficiently (MIYAMOTO et al. 1994) if at all (ZHU et al. 1994), and in vitro translated Res2 does not bind DNA in the absence of Cdc10 (ZHU et al. 1994). Thus it is possible that the DNA binding domain in Cdc10 is functional and contributes to the DNA binding affinity of the Res2/Cdc10 complex.

4.2 Swi6/Cdc10 Repeat Domain

The central domain of Swi6 and Cdc10 is the most conserved region and is found in all the family members. It contains two 33 amino acid repeats which show significant homology to six tandem repeats in the *notch* gene of *Drosophila* (BREEDEN and NASMYTH 1987b). This Swi6/Cdc10 repeat has since been found in over 100 proteins of diverse function (BORK 1993), including membrane proteins, the cytoskeleton component ankyrin, toxins, other transcription factors, and most recently in cyclin-dependent kinase inhibitors (ENDICOTT et al. 1994).

Swi4 (ANDREWS and HERSKOWITZ 1989b) and Mbp1 (KOCH et al. 1993) and their *S. pombe* (TANAKA et al. 1992; CALIGIURI and BEACH 1993; ZHU et al. 1994; MIYAMOTO et al. 1994) and *K.lactis* (KOCH et al. 1993) counterparts also contain two consensus repeats. In addition, the existence of two degenerate repeats between the two consensus repeats, and the N-terminal half of a fifth repeat has been proposed (BORK 1993). Alignment of the eight members of the Swi4/6 family suggests that these degenerate repeats have been conserved through evolution (see Fig. 3). Furthermore, single point mutations have been identified in *cdc10* and *SWI6* that change conserved residues within each of the five repeats and result in protein which is temperature sensitive for function (REYMOND et al. 1992; S.Ewaskow, J.Sidorova and L.Breeden, unpublished). This provides biological evidence that there are not two, but five functional repeats in these proteins. The specific sequences that are conserved in each of the repeats differ substantially, and their individual consensus sequences are shown in Fig. 3. Often the residues differ but the chemical nature of their R groups are conserved. The sequence differences between individual repeats in a given protein may reflect variations in the precise structure or function of each repeat. The fact that many single substitutions in a single repeat have been found that inactivate the protein suggests that each repeat is important for protein function or stability. However, N-terminal truncations of Cdc10 that contain only repeats 4 and 5 can partially complement the temperature sensitive growth defect of *cdc10–129* (AVES et al. 1985). Also, C-terminal truncations that eliminate the fourth and fifth repeat of Res2 still show partial activity (ZHU et al. 1994). Clearly in these cases all five repeats are not required for function. Thus it is possible that a mutant repeat may be more disruptive to the structure and/or function of this domain than a deleted repeat.

The role of the repeats in the Swi4/Swi6 family of transcription factors is not known. They are implicated in protein–protein interaction in other systems (LaMarco et al. 1991; Inoue et al. 1992), but they do not appear to play a role in the interaction between Swi4 and Swi6 (Sidorova and Breeden 1993). Deletion of the entire domain in Swi6 does not prevent its interaction with Swi4 (Andrews and Moore 1992b). Six alanine substitutions in the conserved residues of Swi6/Cdc10 repeats 1 and 4 (underlined in Fig. 3) cause loss of function of Swi6 in vivo but do not impair the ability of Swi4 to coimmunoprecipitate with Swi6 (Sidorova and Breeden 1993). However, these substitutions perturb the ability of the Swi4/Swi6 complex to bind DNA (Sidorova and Breeden 1993). This may indicate a qualitative change in the association between Swi4 and Swi6 which prevents good contact with the DNA. Alternatively, the DNA binding complex may include other unidentified protein(s) required for stable binding and detection on mobility shift gels. If that is the case, the Swi6/Cdc10 repeats may be required for interaction with those proteins. No other specific activators of SCB elements were identified in the screen in which many alleles of *swi4* and *swi6* were identified (Breeden and Nasmyth 1987a). However, it is possible that such an activator could play a more general role in transcription (such as *SWI3*; Breeden and Nasmyth 1987a: Andrews and Herskowitz 1989b), or it could be an essential gene. No conditional screens for SCB activators have ever been performed.

Five different single amino acid substitutions in the Swi6/Cdc10 repeat domain of Cdc10 have been identified that cause temperature sensitivity. As with the Swi6 repeat mutant, these *cdc10* mutant proteins persist in cells, but they show no DNA binding activity (Reymond et al. 1992). It is not yet known whether these ts Cdc10 proteins are still able to bind Res1 and Res2, but it is likely that they can associate because this interaction involves the conserved C-terminal sequences of Cdc10 (see below). If they are able to interact with Res1 and Res2, these mutants will provide further evidence that the Swi6/Cdc10 repeats contribute to the DNA binding activity of the complexes. However, Swi6/Cdc10 repeats are found in many proteins that do not bind DNA, and therefore they are probably affecting DNA binding indirectly.

4.3 Association Domains

Swi4 and Swi6 can be coimmunoprecipitated from cell extracts (Sidorova and Breeden 1993), and coimmunoprecipitations of in vitro translated Swi4 and Swi6 suggest that this is a direct interaction which requires no other yeast proteins (Andrews and Moore 1992b; Primig et al. 1992). The C-terminal 259 residues of Swi4 are sufficient for association with Swi6 (Sidorova and Breeden 1993). The C-terminus of Swi6 is also required for interaction with Swi4 because an 89 amino acid truncation of Swi6 prevents Swi4/Swi6 complex formation (Andrews and Moore 1992b).

The picture is similar in *S. pombe*, where the C-terminal 188 residues of Res1 are sufficient for association with Cdc10. Loss of either the first 50 or the last

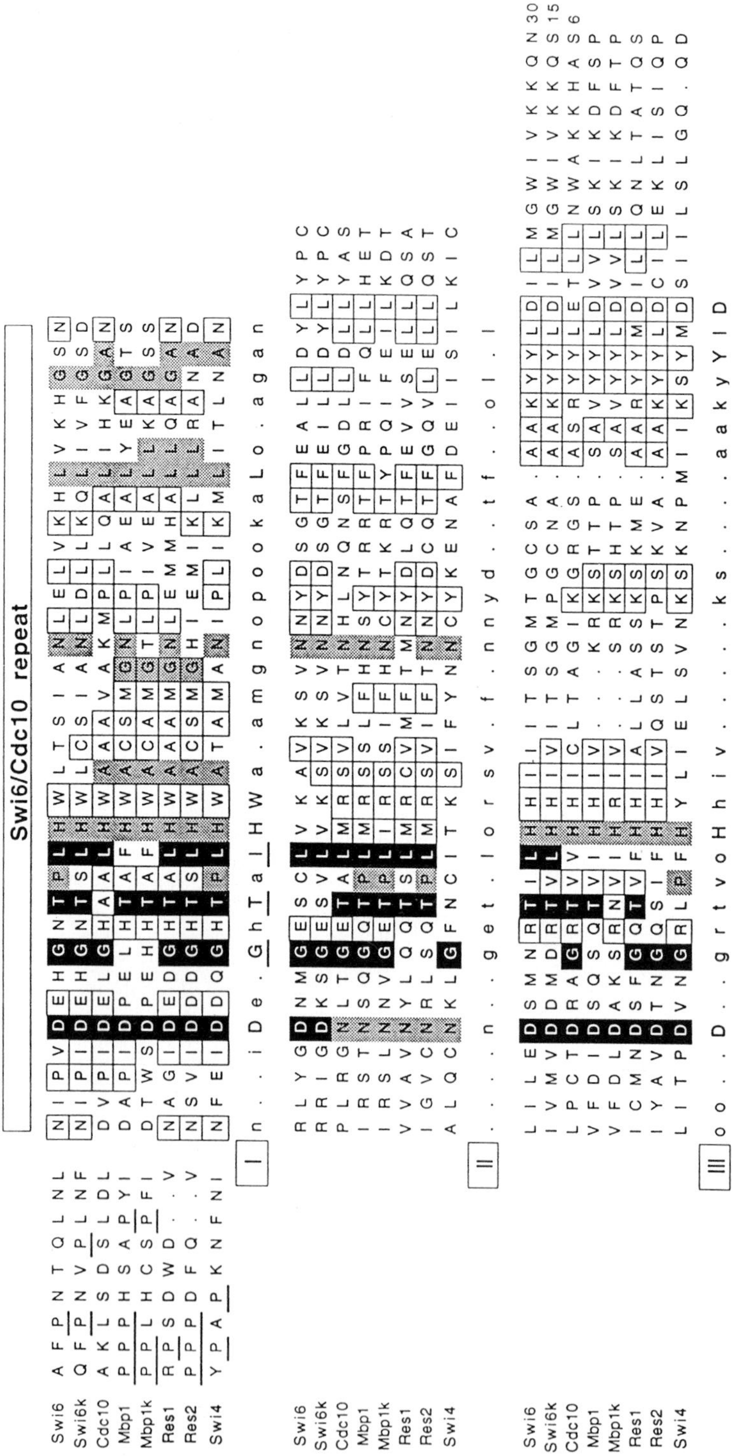

Swi6/Cdc10 repeat
Swi6
Swi6k
Cdc10
Mbp1
Mbp1k
Res1
Res2
Swi4
I
II
III

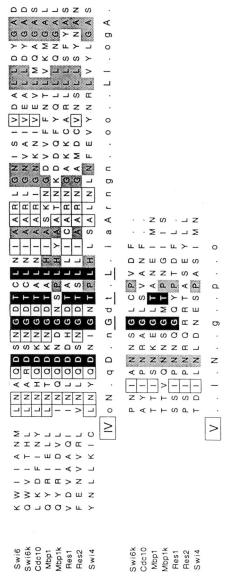

Fig. 3. Alignment of the Swi6/Cdc10 repeat domain shows the residues common to all five repeats (*black*) and the residues conserved in a subset of the repeats (*gray*); *boxed residues*, amino acids conserved in at least four family members but only within a single repeat. The consensus sequence for individual repeats is shown below the alignment. *Upper-case letters*, identify in all eight family members; *lower-case letters*, residues conserved in at least half of sequence; *o*, positions that are always hydrophobic; *underlining*, emphasis on the high propensity for proline residues preceding the Swi6/Cdc10 repeat domain, and, in the consensus sequences of repeats I and IV, the positions of the six alanine substitutions in Swi6 that inactivate Swi6 and disrupt DNA binding. This alignment represents the contiguous sequences of all eight family members except for small nonhomologous inserts that exist in three of the proteins after repeat III. The length of the omitted sequence is noted

48 residues from this region prevents association (Ayte et al., to be published). This suggests either that there is a fairly large domain required for interaction, or that there are at least two separate domains which are required. Several blocks of homology have been identified within the C-terminal region that are conserved in some or all family members. These are shown in Fig. 4. One striking feature is the inferred propensity to form amphipathic α-helices in this region of the proteins, including some that contain heptad repeats of leucine (SIDOROVA and BREEDEN 1993). These are characteristic of leucine zippers, which serve as association domains for many dimeric transcription factors (LANDSCHULZ et al. 1988).

4.4 Nuclear Localization

The proteins in this family that have been analyzed are all nuclear localized for at least part of the cell cycle, but the nuclear localization signal has been defined only in the Swi6 protein of S. cerevisiae. Swi6 is a very acidic protein, and one of its only clusters of basic residues occurs at position 163. This sequence (KKLK) is required for nuclear localization, and its localization is regulated within the cell cycle by phosphorylation of the upstream serine (see Sect. 6.2). This signal is located about 150 residues N-terminal to the Swi6/Cdc10 repeat domain, in a region where there is little or no homology, even between Swi6 and the putative K. lactis Swi6 homolog. Thus, it is unknown whether the nuclear localization signal is located in a similar position or regulated in a similar way in the other proteins.

5 DNA Binding Specificity of the Complexes

The binding specificity of SCB- and MCB-binding complexes has not been analyzed in detail. This is partially due to the fact that the binding complexes are unstable in crude extracts and are likely to be qualitatively different at different times in the cell cycle. Another complication is that the elements are usually found in clusters, and in many cases binding has been detected only using probes containing multiple elements. In vivo a single SCB or MCB element is an extremely weak UAS (BREEDEN and NASMYTH 1987a). These factors complicate the use of site selection (BLACKWELL et al. 1993) or mutagenesis to systematically define a single binding site.

5.1 Swi4/Swi6 Complexes

Specific binding of Swi4 and Swi6 to SCB elements has been demonstrated on HO, CLN2, and PCL1 promoter fragments (ANDREWS and HERSKOWITZ 1989b; OGAS et al. 1991; NASMYTH and DIRICK 1991). However oligomers of MCB elements can compete for this binding in vitro much better than other nonspecific competitors.

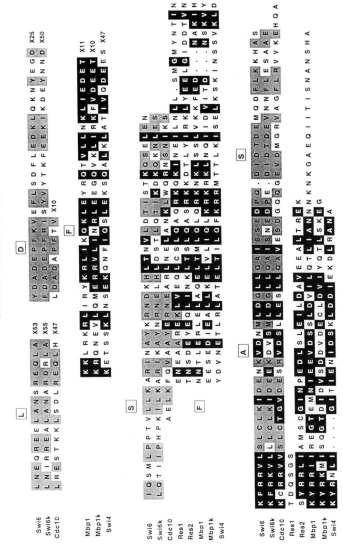

Fig. 4. The terminal regions required for complex formation between Swi6, Cdc10, and their associated proteins. They have been aligned to emphasize regions of high conservation in all members (black-shaded residues in the A block); gray-shaded residues in blocks L, D, and S, regions of homology shared by Swi6, Swi6k, and Cdc10; black-shaded residues in block F, sequences in common between Swi4, Res1, Res2, and the Mbp1 proteins. This alignment represents contiguous sequence. X, Nonhomologous inserts, followed by the number of residues that are not shown

The *CLN1* promoter was thought to be an SCB-regulated gene, because its expression is reduced in *swi4* and *swi6* mutants (OGAS et al. 1991; MOLL et al. 1992). There are no consensus SCBs in this promoter, and a relaxed consensus CNCGAAA for Swi4/6 binding was thus proposed. This relaxed consensus was not supported by mutational analysis, which showed that all seven residues are important (ANDREWS and MOORE 1992a). Furthermore, mutation of the putative SCBs in the *CLN1* promoter has no impact on transcription. There is a cluster of three MCB-like elements 80 base pairs upstream from the putative SCBs that provide most of the UAS activity of this promoter. Despite their resemblance to the MCB consensus, the predominant binding activity in vitro is a complex containing Swi6 and Swi4. Swi4 antibodies, which do not cross-react with Mbp1, supershift the complex, and in vivo *swi4* mutants show a greater defect in transcription from these MCB-like elements than *mbp1* mutants show (Partridge and Breeden, to be published.). The MCB elements from *TMP1*, and the SCBs from the *HO* promoter both effectively compete for the *CLN1* binding activity.

It is not clear whether these differences in the binding site preference of the Swi4/Swi6 complexes reflect differences in their subunit composition at each site, or whether flanking DNA sequences or proteins bound to adjacent sites on the DNA are involved. In vitro translated fragments of the N-terminal domain of Swi4 can bind both SCB and MCB sequences (PRIMIG et al. 1992). Carboxyethylation interference indicates that this Swi4 N-terminal fragment binds to a 10-bp stretch in the major groove of the *CLN2* promoter which includes three bases (GTA) 5' to the CACGAAA sequence (PRIMIG et al. 1992). These upstream sequences (see Fig. 5) do not conform to the extended consensus originally proposed based on conservation between sites in the *HO* promoter (Pur NN Pyr CACGAAAA; NASMYTH 1985b), and they look very similar to the interference pattern of an MCB binding complex on the *TMP1* promoter (compare boxed residues in Fig. 5). Swi4 and Swi6, translated in vitro in reticulocyte lysates, can also form a complex on SCB elements from the *CLN2* promoter, but this complex migrates slightly faster than the complex isolated from yeast cell extracts (PRIMIG et al. 1992). This could indicate the presence of other proteins in the complex or changes in their modification states.

5.2 Mbp1/Swi6 Complexes

MCB binding on a *TMP1* promoter fragment occurs over two oppositely oriented MCB elements, in the major groove of the DNA, and requires two additional A-T base pairs 3' to the T of the Mlu1 site (MOLL et al. 1992; see Fig. 5). At least one of these A-T pairs is conserved in most MCB elements (JOHNSTON and LOWNDES 1992). This gives these binding sites considerable similarity to the E2F binding site of higher cells (MEAN et al. 1992) and to the SCB elements of *CLN2*. The *TMP1* binding activity can be competed away by either SCB- or MCB-containing DNA (MOLL et al. 1992), and the binding activity is reduced in extracts from *swi4* cells (KOCH et al. 1993). This suggests that the binding complex observed on the *TMP1*

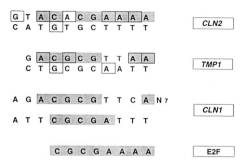

Fig. 5. Similarities between the known cell cycle regulatory promoter elements in yeast and higher cells. The *CLN2* SCB element and *TMP1* MCB elements are shown as duplex DNA; *boxed residues*, the positions at which carboxyethylation interferes with DNA binding (PRIMIG et al. 1992; MOLL et al. 1992). These binding sites are thought to be bound by Swi4/Swi6 and Mbp1/Swi6, respectively, and their interference pattern is quite similar. Two adjacent regulatory elements from the *CLN1* promoter are also shown. These sites look more like MCB elements, but they are primarily bound by Swi4/Swi6 complexes (Partridge and Breeden, to be published). Below, for comparison, the E2F/DP1 binding site which promotes S phase specific transcription in higher cells is shown (MEAN et al. 1992). Shaded residues represent sequence similarities

elements could be a mixture of Mbp1/Swi6 and Swi4/Swi6 complexes. In vivo the distal MCB is most critical for transcription activation. Mutations in this site decrease transcription five-fold (MCINTOSH et al. 1991), and this drop occurs in both wild-type and *mbp1* deletion strains (KOCH et al. 1993). Clearly Mbp1 is not the only transcription factor that activates transcription from this site, and Swi4 is the most likely candidate. The similarity of the carboxyethylation interference pattern on the *CLN2* SCB and the MCB elements of *TMP1* indicate that the complexes that form on these two sites have very similar points of DNA contact despite the differences in the sequences.

The *CLB5* promoter is another example where Swi4 and Mbp1 may both be regulating transcription from MCB elements. *CLB5* is a cell cycle regulated cyclin gene, with five potential MCB elements located upstream from its open reading frame (EPSTEIN and CROSS 1992; SCHWOB and NASMYTH 1993). Although *CLB5* transcription has not yet been shown to be mediated by these MCB elements, *CLB5* mRNA peaks at the G_1/S boundary and behaves as *TMP1* transcription in the arrest caused by a G_2 cyclin deficiency (AMON et al. 1993). This led to the proposal that *CLB5* is an Mbp1/Swi6 regulated gene. However, *CLB5* continues to be cell cycle regulated in both *mbp1* and *swi4* mutant strains (EPSTEIN and CROSS 1992; KOCH et al. 1993). This is most consistent with the view that both the Swi4/Swi6 and Mbp1/Swi6 complexes can form on the *CLB5* promoter and regulate its transcription. This is in contrast to the *HO* promoter, which is absolutely dependent upon Swi4 (BREEDEN and NASMYTH 1987a). In the context of the *HO* promoter Mbp1 has no compensating activity for Swi4.

All these studies indicate that the preferred binding sites for Swi4/Swi6 and Mbp1/Swi6 cannot be predicted based on our current understanding of the SCB and MCB consensus sequences. It is possible that flanking sequences or associations with adjacent DNA binding proteins are important for stabilizing these different complexes. It is also possible that Mbp1 and Swi4 have largely redundant DNA binding activities, and that there are other associated proteins which confer the differences between the complexes. There are many genes that are regulated within the cell cycle such that their transcripts accumulate specifically in late G_1–early S phase and that have MCB elements in their promoters. In most of these cases the role of MCB elements has not been determined, nor have the binding proteins been defined. Based upon the promoters that have been studied in detail, it is perhaps premature if not wholly incorrect to assume that all these genes are regulated by Mbp1 and Swi6. At this point it is also premature to view the MCB binding complex and the SCB binding complex as particularly distinct transcription complexes or to assume that there are only two such complexes.

5.3 Cdc10 Complexes with Res1 and Res2

Many fewer cell cycle regulated genes have been identified in *S. pombe*, but the story is already similar to that described for *S. cerevisiae*. In the *cdc22* promoter an MCB-like element is bound by Cdc10/Res1 and Cdc10/Res2 complexes (CALIGIURI and BEACH 1993; ZHU et al. 1994). In the *cdt1* promoter an apparently unrelated sequence is bound by Cdc10 (ATAACGATGCAT). Binding to the novel site in the *cdt1* promoter can be competed by MCB sequences but not by mutant MCBs (HOFMANN and BEACH 1994). The functional significance of the novel binding site and Cdc10's binding partner at this site have not been reported. Interestingly, the *cdt1* upstream region also contains a cluster of three MCB-like sequences very near the putative start site for translation. The position of these MCBs with respect to the mRNA start site has not been determined, nor has the *bona fide* translational start been identified; therefore it is possible that this cluster of MCBs may contribute to the cell cycle regulation of *cdt1*. Nevertheless, the binding of Cdc10 to the novel upstream site may lead to the identification of a new binding partner for Cdc10, or it may lead to a second consensus sequence for Cdc10/Res1 binding.

Res2 is the newest member of the *S. pombe* MCB binding proteins. It is unique in that it seems to bind specifically to MCB elements. Its binding to the *cdc18* promoter cannot be competed by the SCB elements from *HO* or the E2F binding sites from the adenovirus E2A promoter (ZHU et al. 1994). Interestingly, Res2 has the most limited homology in the DNA binding domain and the largest requirement for association with Cdc10 for stable DNA binding activity (ZHU et al. 1994; MIYAMOTO et al. 1994).

6 Cell Cycle Regulation of Transcriptional Activity

The subject that has received the most attention with regard to this family of transcription factors is the cell cycle regulation of their activities, particularly in the context of G_1 cyclin transcription. G_1 cyclins are required for the G_1 to S transition (RICHARDSON et al. 1989). Most of the G_1 cyclins are transcriptionally regulated and their mRNAs peak at about the time of the G_1 to S transition. Dominant mutations that stabilize *CLN3* or *CLN2* speed the transition to S phase (CROSS 1988; NASH et al. 1988), and constitutive overproduction of *CLN2* causes premature entry into S phase and death (AMON et al. 1993). Constitutive overproduction of *CLB5*, a putative MCB-regulated cyclin involved in S phase regulation, prevents cell cycle arrest in response to mating pheromones (EPSTEIN and CROSS 1992; SCHWOB and NASMYTH 1993). These results suggest that the level of cyclins expressed in late G_1 determines when and whether the G_1 to S transition occurs. If this is the case, the machinery that regulate cyclin transcription must respond to both extra- and intracellular signals to modulate cyclin transcription, and thus control progression through the cell cycle.

The initial characterization of the G_1 cyclin promoters *CLN1*, *CLN2* and *PCL1* was carried out before the MCB element was identified, and thus only the SCB and SCB-like sequences were noted as potential regulators (OGAS et al. 1991). Later, when deletion analysis was carried out, it became clear that MCB-like elements are also present and active in the *CLN1* and *CLN2* promoters, and that the SCB-like sequences do not activate *CLN1* transcription at all. In addition, both the *CLN1* and *CLN2* promoters contain at least one other cell cycle regulated element that has not yet been identified (STUART and WITTENBERG 1994; CROSS et al. 1994; Partridge and Breeden, to be published), and there are at least three independent pathways of *trans*-activation that contribute to *CLN1*, *CLN2* and *HO* transcription (BREEDEN and MIKESELL 1994). This makes it clear that regulation of G_1 cyclin transcription is much more complicated than originally thought. Nevertheless, understanding the cell cycle-specific activation of SCB- and MCB-mediated transcription is a critical step in determining what controls the G_1 to S transition in *S. cerevisiae*.

6.1 The Role of Cdc28 in Activation

SCB elements are inactive in cells arrested in G_1 with a *cdc28* mutation (BREEDEN and NASMYTH 1985, 1987a). MCB-regulated transcripts are also reduced in *cdc28* arrest (JOHNSTON and THOMAS 1982). This could reflect a direct activation of the transcription complex by the Cdc28 kinase, or it could be indirect, owing to the fact that *cdc28* mutants arrest at a stage in the cycle when SCB and MCB elements are not yet active. Most events in the mitotic cycle follow, and depend upon, the execution of the Cdc28-dependent step(s) in G_1; it is only a question of how many steps there are between the two events. Nevertheless, activation of SCB- and

MCB-mediated transcription occurs very shortly after Cdc28 activation in G_1 and is likely to be more directly affected than are the later events.

The proposal that SCBs are responsible for G_1 cyclin transcription led to two studies (CROSS and TINKELENBERG 1991; DIRICK and NASMYTH 1991) that showed that inactivation of Cdc28, either by using a temperature-sensitive allele of *cdc28* or by elimination of the three G_1 cyclins known to activate Cdc28, prevent peak transcription of *CLN1* and *CLN2*. This supported the view that SCB's or a similarly regulated pathway is important for G_1 cyclin transcription. It also led to the formulation of the positive feedback loop model (See Fig. 6). The G_1 cyclins are required for the maximal transcription of the G_1 cyclins. Thus by definition there is a positive feedback mechanism. In its simplest form the proposal is that a constitutive cyclin, perhaps Cln3 (TYERS et al. 1993), binds and activates Cdc28. The active Cdc28/Cln3 complex phosphorylates and activates the Swi4/Swi6 complex which leads to expression of *CLN1* and *CLN2*. The Cln1 and Cln2 cyclins can then bind and activate more Cdc28 which activates more Swi4/Swi6 complexes. This leads to a rapid increase in *CLN* transcription and peak expression of many of the gene products involved in DNA synthesis. This is an appealing model because a positive feedback loop would cause a concerted and irreversible transition to S phase.

One piece of evidence which supports this model is the finding that reactivation of a temperature labile Cdc28 induces transcription of several SCB- and MCB-regulated genes even under conditions that prevent new protein synthesis (MARINI and REED 1992). Although it is difficult to assess the magnitude of the induction in these experiments, they suggest that new protein synthesis is not required for some MCB and SCB activation. Reactivating the kinase leads either directly, or through a posttranslational regulatory cascade, to increased transcription of these cell cycle regulated promoters.

6.2 Cell Cycle Regulated Phosphorylation and Nuclear Localization of Swi6

The experiments described above led to a detailed analysis of phosphorylation of Swi6, in which it was found that serine 160 of Swi6 displays cyclic phosphorylation (Sidorova and Breeden, to be published). Its phosphorylation peaks before S phase and remains high for most of the cycle. During late nuclear division the phosphorylation level declines and remains low throughout most of the next G_1 phase. Ser-160 resides within a Cdc28 consensus phosphorylation site (MORENO and NURSE 1990) and is followed by lysine residues (SPLKKLK). These lysines are required for nuclear localization of Swi6, and its movement to the nucleus is correlated with hypophosphorylation of Ser-160 (Sidorova and Breeden, to be published.). Swi6 is predominantly cytoplasmic throughout the period of the cell cycle during which it is phosphorylated. Swi6 is detectable in the nucleus only very late in mitosis and throughout G_1 (TABA et al. 1991; Sidorova and Breeden, to be published), which is coincident with the interval during which Ser-160

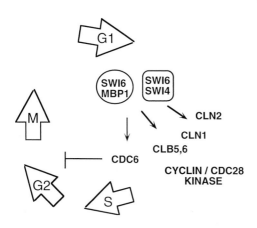

Fig. 6. Start-specific transcription in *S. cerevisiae*. The Swi4/Swi6 and Mbp1/Swi6 transcription complexes activate transcription of dozens of genes at the G_1 to S transition. They are required for the first wave of cyclin transcription. These cyclins associate with and activate the Cdc28 kinase, which starts the mitotic cycle. They are also involved in the transcriptional control of Cdc6, which is a protein required for initiation of DNA synthesis, and which may also be involved in delaying mitosis until S phase is complete

phosphorylation is greatly reduced. Substitution of Ser-160 for alanine results in constitutive nuclear localization. This suggests that phosphorylation of Ser-160 prevents nuclear entry. Aspartic acid substitutions at 160 cause Swi6 to persist predominantly in the cytoplasm. The simplest explanation of this result is that the negatively charged aspartic acid phenocopies the negatively charged phosphoserine, and also impairs nuclear entry. This sort of phosphorylation-mediated regulation of nuclear localization has been observed in many other proteins (NASMYTH et al. 1990; HUNTER and KARIN 1992) and may simply be a matter of masking or neutralizing the positive charge of adjacent nuclear localization sites.

The kinase responsible for Ser-160 phosphorylation has not been identified, but Cdc28 (MENDENHALL et al. 1987), Pho85 (KAFFMAN et al. 1994), Kin28 (SIMMON et al. 1986) and Spk1 (ZHENG et al. 1993) are likely candidates. Ser-160 is hypophosphorylated in a *cdc28-13* arrest (Sidorova and Breeden, to be published), but whether this is due to the loss of Cdc28 kinase activity or is the result of halting the cell cycle during G_1 (when Ser-160 is normally hypophosphorylated) is not known. Despite the fact that Swi6 phosphorylation occurs at about the same time as the pulse of SCB- and MCB-regulated transcription, neither the alanine nor the aspartic acid substitutions at position 160 in Swi6 have a major impact on the cell cycle regulation of target genes (Sidorova and Breeden, to be published). Furthermore, alanine substitution at all four of the potential sites for Cdc28 phosphorylation has no appreciable impact on transcription in synchronously cycling cells. Thus, phosphorylation at these sites is not required to activate Swi6 in the SCB or MCB transcription complex.

It is possible that Swi4 or another unidentified protein in the SCB and MCB transcription complex is regulated by phosphorylation. Another intriguing possibility is that the phosphorylation is not important at all. Perhaps Swi6 is phosphorylated because it is bound to a kinase, but the importance of the kinase interaction is that its binding to the Swi4/Swi6 complexes serves to localize the kinase to particular cell cycle regulated promoters. Perhaps the kinase, thus positioned at these promoters, phosphorylates and activates proteins in the general transcription complex and induces transcription. The kin28 (MO15) kinase can be found associated with the general transcription machinery in yeast (FEAVER et al. 1994) and in higher cells (ROY et al. 1994). Kin28 is related to Cdc28 (SIMON et al. 1986) and is associated with a cyclin-like protein (VALAY et al. 1993). It phosphorylates the C-terminal domain (CTD) of the largest subunit of RNA polymerase, which is composed of a heptapeptide repeat with the consensus sequence of YSPTSPS. The CTD is required for initiation of transcription at nearly all *Pol*II transcribed genes, and its phosphorylation may play a role in facilitating elongation of the initiated transcripts (for review see DAHMUS 1994). The MO15 kinase is also required to activate cyclin-dependent kinases in higher cells (POON et al. 1993; FESQUET et al. 1993). If this function is conserved in *S. cerevisiae*, then another possibility is that the Swi4/Swi6 complex delivers Cdc28 to the promoter, at which Cdc28 is phosphorylated and activated by Kin28. Cdc28 leaves the complex to phosphorylate other nuclear substrates, and transcription ensues.

6.3 Swi4 Transcription Is Cell Cycle Regulated

Swi4, which is a critical component of the Start-specific transcription machinery is also cell cycle regulated at the transcription level (BREEDEN and MIKESELL 1991). The Swi4 promoter contains a cluster of three MCB-like elements that provide 90% of the promoter's transcriptional activity. In addition, it contains at least one other element that confers cell cycle regulation. This novel element activates *SWI4* transcription 5–10 min earlier than MCB and SCB-mediated transcription (FOSTER et al. 1993) and is probably responsible for the earlier timing of *SWI4* transcription compared to SCB or MCB-mediated transcription in wild-type cells. The early timing of Swi4 transcription is consistent with its role as an activator of *CLN* transcription and the activation of this early cell cycle box may play a crucial role in initiating the regulatory cascade that promotes the G_1 to S transition.

When *SWI4* is expressed throughout the cell cycle, the cell cycle regulation of Swi4's target promoters (*HO, CLN1*, and *CLN2*) is reduced but not eliminated (BREEDEN and MIKESELL 1991, 1994). There is a two- to tenfold increase in transcript accumulation during periods of minimum transcription, but periodicity is still observed. This indicates that the restricted expression of *SWI4* contributes to the regulation of these promoters, but it is not the only source of control. Something else must be preventing high level transcription from these promoters, despite the fact that their primary activator is highly overproduced. Both deletion of Swi6 and overproduction of Swi4 almost eliminates the cell cycle regulation of *CLN1* and *CLN2* (BREEDEN and MIKESELL 1994). This is consistent with the proposal that Swi6 is both a negative and positive regulator (see Sect. 6.5). However, even under these conditions, a twofold fluctuation in *CLN* transcription can still be observed. There must be a third level of regulation, independent of the presence of Swi6 or the periodic expression of Swi4 that also activates transcription of these promoters during the G_1/S interval (BREEDEN and MIKESELL 1994).

The *Res2* transcript of *S. pombe* is also slightly induced during G_1 and S compared to the rest of the cycle (OBARA-ISHIHARA and OKAYAMA 1994), but the significance of this regulation has not been investigated. In contrast, neither Res1 (OBARA-ISHIHARA and OKAYAMA 1994) nor Mbp1 (KOCH et al. 1993) are regulated during the cell cycle at the level of transcription.

6.4 Complex Formation During the Cell Cycle

Despite the fact that Swi6 is concentrated in the nucleus during only a fraction of the cell cycle, and the transcription of *SWI4* is also cell cycle regulated, there is no evidence from in vitro studies that the association between Swi4 and Swi6 or their binding to the SCB elements varies within the cell cycle. This may be due to the use of whole cell extracts in these studies and/or the intrinsic difficulties in obtaining highly synchronous cultures. One mobility shift change has been observed in the SCB binding complex during recovery from either α-factor or elutriation, but this shift is not coincident with Start, or with the interval of *HO* activation or repression (TABA et al. 1991). The probe used to observe the novel complex contained two perfect SCB elements, and therefore it is possible that the observed shift to a larger size represents the binding of two SCB complexes instead of one.

Several in vitro binding studies carried out with MCB elements and crude cell extracts indicate that MCB binding varies within the cell cycle, both in *S. cerevisiae* and in *S. pombe*. In vitro binding to three tandem MCB elements varies in synchronized cell extracts of *S. cerevisiae* (LOWNDES et al. 1991), and this binding mirrors the timing of Swi6 localization to the nucleus. The Cdc10 protein of *S. pombe* is detectable in the nucleus throughout the cell cycle, but the level of Cdc10 in the nucleus is reduced during anaphase and G_1 (REYMOND et al. 1993). This reduction in nuclear localization of Cdc10 also coincides with the reduction of MCB binding activity, as assayed by binding studies using synchronized cell extracts and the *cdc22* promoter (REYMOND et al. 1993). It is interesting to note that the nuclear localization of both Swi6 and Cdc10 is cell cycle regulated, but the interval of time during which they are in the nucleus is quite different. Swi6 is cytoplasmic throughout S and G_2/M, and becomes localized to the nucleus very late in the cell cycle and throughout G_1, while Cdc10 shows the opposite pattern of localization. The significance of this regulated localization is not understood. Presumably both proteins are predominantly nuclear at the G_1 to S transition when they are active as transcription factors, but at least in the case of Swi6, mutations which cause constitutive nuclear localization have no impact on the regulation of SCB- or MCB-mediated transcription (see Sect. 6.2).

6.5 Negative Control

Swi6 is the common component of the MCB and SCB complexes. It is thought to play both a negative and a positive role in MCB transcription because *swi6* mutants express intermediate levels of *TMP1, RNR1, CDC9* (LOWNDES et al. 1992b; DIRICK et al. 1992) and *SWI4* (FOSTER et al. 1993) constitutively through the cell cycle. Thus, Swi6 is required for both the full activation of transcription at the G_1 to S transition and for full repression during the rest of the cycle. Apparently there is an independent path of activating MCB elements which is active throughout the cell cycle. When Swi6 and Mbp1 are present, the other factors cannot interfere, and normal periodic transcription ensues. When either protein is absent, MCB-regulated promoters are still transcribed by this alternative constitutive pathway. This helps to explain why *SWI6* and *MBP1* are not essential genes. However, if this is correct, at least some Swi6 must remain in the nucleus of wild-type cells throughout the cell cycle. DIRICK et al. (1992) have observed complexes on MCB elements throughout cell cycle, and the differences between these studies and those that demonstrated periodic binding have not been elucidated.

Swi6 may also have a negative role in SCB-mediated transcription because Swi6 activity is correlated with the repression of *HO* transcription. This is particularly evident during the G_1 arrest caused by the *cdc28-4* mutation (BREEDEN and MIKESELL 1994). For example, Swi4 overproduction does not eliminate G_1 repression; *HO* is still repressed in a *cdc28* arrest, as it is in wild-type cells. However, when Swi4 is overproduced in a *swi6* mutant strain, *HO* is deregulated and expressed at a high level during a *cdc28* arrest. The same deregulation occurs in cells carrying a C-terminal truncation of Swi4. In this case the Swi6-association domain of Swi4 is absent, and Swi6 is thus not a part of the complexes that form on SCB elements. These complexes activate transcription of *HO*, but it is deregulated and expressed during G_1. The simplest explanation of this is that Swi6 is the target of negative regulation during G_1. Its presence in complexes with Swi4 on target promoters prevents their transcription during G_1. After activation of the Cdc28 kinase the SCB-binding complexes are modified in some way that lifts repression and allows their transient activation, which is associated with Start.

Swi4 is the primary activator at SCB elements. Swi4 contains the DNA binding domain, and when it is overproduced, it can activate transcription of *HO*, *CLN1* and *CLN2* in the absence of Swi6. Thus, one way to prevent SCB-mediated transcription would be to sequester or inactivate Swi4. AMON et al. (1993) have shown that *CLN1, CLN2* and *PCL1* transcription remains high in cells that are arrested in G_2 due to the loss of four G_2 cyclins (Clb1-Clb4). In contrast, the *POL1, CLB5* and *RNR1* promoters, which are thought to be regulated by Mbp1 and Swi6, turn off in this G_2 arrest. This observation distinguishes Swi4/Swi6-regulated transcription from that of Mbp1/Swi6 and gives rise to the hypothesis that the G_2 cyclins play a role in inactivating Swi4. These authors also found that Swi4 binds and can be phosphorylated by Clb2/Cdc28 kinase, and suggested that this could be the mechanism for its inactivation. However, the fact that Swi4/Swi6 complexes on SCBs can be observed throughout the cell cycle (TABA et al. 1991), and *CLN1* and *CLN2* transcription is still periodic in a *swi4* mutant (STUART and WITTENBERG 1994; CROSS et al. 1994; BREEDEN and MIKESELL 1994) indicates that the binding of Swi4 to Clb2 and its proposed inactivation of Swi4 are not required to turn off *CLN* transcription. If this is a source of negative control, it must be redundant with some other negative control pathway.

7 Importance of Start-Specific Transcription for Cell Cycle Progression

Cdc28 kinase activity depends upon its association with a cyclin and control of this kinase activity controls cell cycle progression. Nine of the ten known cyclins of *S. cerevisiae* have been shown to be periodically expressed, and in the studies that have been carried out the control was exerted at the level of transcription initiation. The wave of regulation that is MCB- and SCB-dependent affects the G_1 and S phase specific cyclins and occurs at the G_1 to S transition, but there are two other kinetically distinguishable waves of transcription that control Clb3-Clb4 and Clb1-Clb2 transcription, respectively (FITCH et al. 1992; RICHARDSON et al. 1992). It is difficult to imagine that this complex pattern of transcriptional control throughout the cell cycle serves no purpose. However, the standard experiment to establish the importance of cell cycle regulated transcription of a gene, constitutively transcribing it in vivo and monitoring the phenotype, has yielded few compelling examples of its importance.

   CLN2 transcription, driven constitutively by the *GAL1* promoter, is apparently lethal to yeast cells (AMON et al. 1993). This indicates that the overproduction and/ or the constitutive expression of Cln2 is deleterious, but the nature of the arrest has not been fully characterized. When overproduction is involved, it is possible that the deleterious effect is due to phosphorylation of a protein that would not be a good substrate under physiological conditions. In contrast, *CLN1* can be constitutively transcribed from the *GAL1* promoter with few apparent effects (RICHARDSON et al. 1989). This has led to the use of the inducible *GAL: CLN1* to obtain a synchronous start of the cell cycle in *CLN*-deficient strains in order to study other cell cycle regulated events. Whether this is a prudent synchronization strategy, and whether a physiologically relevant cell cycle ensues under these conditions are questions that await a better understanding of the control of Cln1 activity. Nevertheless, the fact that providing one Cln protein constitutively in a *cln1 cln2 cln3* mutant background does not disrupt the orderly process of cell division casts doubt on the importance of the Start-specific pulse of cyclin transcription. However, this strain (*cln1 cln2 cln3 GAL: CLN1*) still contains at least four other cyclins: Pcl1 (Hcs26), Pcl2 (OrfD), Clb5, and Clb6, which are specifically expressed at the G_1/S border (OGAS et al. 1991; TYERS et al. 1993; EPSTEIN and CROSS 1992; SCHWOB and NASMYTH 1993), and that are genetically redundant with Cln1 and Cln2 in some contexts (EPSTEIN and CROSS 1992; ESPINOZA et al. 1994; MEASDAY et al. 1994). These cyclins would still be providing a pulse of kinase activity, if such a pulse is required.

Another complicating factor in discerning the importance of transcriptional regulation is the possibility of overlapping levels of control. For example, the CLB5 promoter is transcribed specifically at the G_1 to S transition, and the Clb5 cyclin has a destruction box, which targets it for degradation (GHIARA et al. 1991). This should lead to the transient expression of Clb5 and a pulse of Clb5/Cdc28 kinase activity. However, in addition to this regulation, there is an inhibitor of the Clb5/Cdc28 kinase, called Sic1 (MENDENHALL 1993). Sic1 is also transcriptionally regulated within the cell cycle so that it is present throughout G_1 (DONOVAN et al. 1994). At or around the transition to S phase Sic1 is degraded by the Cdc34-dependent ubiquitin conjugating pathway and the Clb5/Cdc28 kinase activity is restored (SCHWOB et al. 1994). Thus, there are at least four levels of regulation exerted upon the activity of the Clb5/Cdc28 kinase, and this is not necessarily a comprehensive list. In this context it is not surprising that cells that transcribe the CLB5 gene at a high constitutive level grow well in culture. However, GAL: CLB5 containing cells cannot arrest their cell cycle in response to the mating pheromone α-factor (SCHWOB and NASMYTH 1993). This is probably because the Clb5/Cdc28 kinase is not inhibited by Far1, which is another inhibitor that specifically inhibits the G_1 cyclin/Cdc28 complexes (PETER and HERSKOWITZ 1994).

CDC6 is another G_1/S-regulated gene whose constitutive overproduction has a deleterious effect on the cell cycle. CDC6 is not a cyclin. It is one of the many genes required for DNA synthesis (HARTWELL 1976) which are regulated, in part, by MCB elements (ZHOU and JONG 1993). However, Cdc6 is unique in that it appears to have two roles in the cell cycle, one in the initiation of DNA synthesis and the other in preventing mitosis until S phase is complete. As such it may be the best example of a pivotal control protein, whose expression must be tightly controlled for orderly progression through the cell cycle. Cdc6 was identified as an S. cerevisiae activity that could suppress the lethal, premature entry into M phase in S. pombe caused by overproduction of Cdc25 or by hyperactivation of Cdc2. Later it was shown that GAL1-induced transcription of CDC6 also delays mitosis in wild-type S. cerevisiae cells (BUENO and RUSSELL 1992, see Fig. 6). The closest known relative to Cdc6 in S. pombe is cdc18. Interestingly, it is also transcribed at the beginning of the cell cycle by Cdc10-containing transcription complexes, and has similar activities (KELLY et al. 1993).

The fourth example of a gene that is transcribed during the G_1/S interval, and this has a deleterious effect when constitutively overproduced is the Rad53/Spk1/Mec2/Sad1 kinase. The promoter of this gene is uncharacterized, but it contains MCB-like elements and is transcribed at the same time as CLN1 (ZHENG et al. 1993). Mutations in this gene cause rapid death in response to UV irradiation, hydroxyurea (ALLEN et al. 1994), and cdc13 arrest (WEINERT et al. 1994). Since these mutants do not lose viability during arrests in G_1 due to α factor (ALLEN et al. 1994) or cdc28 arrest (WEINERT et al. 1994), it is clearly an activity required during S phase and for the normal arrest in response to DNA damage. Although the phenotype of the rad53 defect has not been characterized, it is clear that constitutive overproduction of the Rad53 kinase slows cellular growth (ZHENG et al. 1993).

8 Conclusion

The two most interesting questions remain unanswered. How are the activities of the Swi4/6 family of transcription factors regulated within the cycle, and what part does their regulated activity play in controlling the G_1 to S transition? One conclusion, which is already evident, is that there is not a simple answer to either question. The complexes have not yet been fully defined, and until they are, changes in these complexes during the cell cycle cannot be fully examined. Nevertheless, several levels of variation through the cell cycle have been observed that could be important for cell cycle regulation. However, elimination of any one source of regulation has little effect on the regulation of the target genes. This indicates that the cell cycle-dependent changes that have been observed are either irrelevant, or that their importance is being masked by redundant pathways of regulation.

Redundant regulatory networks are a prudent strategy for regulating important processes, because multiple mutations are required to fully disrupt control. They are also advantageous because they increase the number of targets amenable to regulation by the intra- and extracellular signals that modulate cell division and coordinate it with other cellular processes. For a colonial micro-organism, which spends most of its time in G_0, perhaps it makes most sense to have a default repression system, with multiple ways to activate the cycle in response to external cues. At least in *S. cerevisiae* this activation results in the rapid and transient induction of the transcription of dozens of genes. Perhaps many, if not most, of these genes could be expressed constitutively without any negative impact, but some, such as *CLN2, CLB5*, and *CDC6*, cannot be continually expressed without serious repercussions. Yeast cells had to evolve affective means to repress these transcripts after the cycle is initiated, no matter how they were activated, or how strong the stimulus. Redundant levels of regulation, or at least the ability to compensate for hyperactivation at one step is not so surprising in this context.

Redundancy has been documented at several levels in this system. One of the primary levels of redundacy is at the promoter level. It is now clear that the promoters of *CLN2* (STUART and WITTENBERG 1994; CROSS et al. 1994), *SWI4* (FOSTER et al. 1993), and *CLN1* (Partridge and Breeden, to be published) contain multiple cell cycle regulatory elements, which are activated at about the same time in the mitotic cycle, but are probably regulated by different mechanisms. After systematic elimination of the 13 known cell cycle regulatory elements in these three promoters, their transcripts continue to be cell cycle regulated by promoter elements that have not yet been identified. These novel elements must be located and characterized before the properties of the composite promoters can be interpreted.

Another level of redundancy that is evident in these regulatory pathways is at the protein level. Swi4/Swi6 and Mbp1/Swi6 complexes bind to both MCB and SCB elements in vitro. Mobility shift experiments indicate that a single complex predominates on any one site, but if this complex cannot be formed, it is likely that the other complex will bind instead. In addition, at least one other transcription factor must exist that can constitutively activate MCB elements when Swi6 is absent. These redundant activities provide a sound survival strategy for the yeast, but they make simple interpretation of mutant phenotypes impossible. For example, exponentially growing *Swi6* mutant strains produce high levels of MCB-regulated transcripts; thus, initially there was no indication that Swi6 was involved in their transcription. Only after it was shown that the Swi6 protein binds to MCB elements was it discovered that MCB-regulated transcripts are deregulated within the cell cycle in *Swi6* mutant strains (LOWNDES et al. 1992b; DIRICK et al. 1992).

For progress to be achieved the complexity of the problem must be reduced. Initially all the promoter elements will have to be studied in isolation. The elements most amenable to study are those that are bound by only one transcription complex, and that are active as single elements. However, the sequence requirements that specify preferential binding by one complex are not yet known. *HO* is the only tightly Swi4- and Swi6-dependent gene known, and therefore it is the most likely to contain such elements. However, there are at least eight SCB elements in the *HO* promoter, and it is not known which of these are active (NASMYTH 1985b). Single SCB and MCB sites, synthesized based on the current consensus sequence, show very little UAS activity in vivo in the contexts in which they have been tested (BREEDEN and NASMYTH 1987a; LOWNDES et al. 1992a). Oligomerization of these elements is one solution, but the behavior of these artificial constructs may not reflect the behavior of "real" promoters. Only when all the regulatory components of a single promoter have been identified will it be possible to study them systematically in their normal contexts and to make real progress in our understanding of cell cycle regulated transcription.

Redundancies and unknowns also limit our ability to determine the importance of the G_1/S-specific transcription in progression of the mitotic cycle. Clearly, the artificial environment of the laboratory, where cells are grown in liquid culture under optimal conditions, has limited the analysis of the fitness, adaptability, and survival of these strains and has skewed our understanding of the cell cycle. However, in addition, redundant posttranscriptional regulatory steps often exist which make changes in transcription regulation phenotypically subtle.

In the early days of yeast molecular biology, if one deleted a yeast gene and found it to be a lethal event, the difficulty of studying an essential gene had to be confronted, but it was reassuring to know that it was an important gene. More recently, particularly in the field of signal transduction, if an activity is important, it is likely to be encoded by two genes. In this case the difficulty of lethality is still confronted with the double mutant, and there is the additional ambiguity over the extent to which the activities overlap. If the lessons of cell cycle regulation can be generalized, one would have to say that if an activity really has to be controlled, there will be at least two independent circuits dedicated to accomplishing that control.

Acknowledgments. I would like to thank B. Andrews, I. Herskowitz, J. Sidorova, and J. Partridge for their helpful comments on the manuscript. I would also like to thank J. DeCaprio, L. Herrington, B. Andrews, M. Ward, S. Garrett, J. Sidorova, and J. Partridge for communicating their unpublished results. Work in this laboratory is funded by grants from NIH GM41073 and the DOD Breast Cancer Initiative DAMD17-94-J-4122.

References

Allen JB, Zhou Z, Siede W, Friedberg EC, Elledge SJ (1994) The *SAD1/RAD53* protein kinase controls multips checkpoints and DNA damage-induced transcription in yeast. Genes Dev 8: 2416–2428

Amon A, Tyers M, Futcher B, Nasmyth K (1993) Mechanisms that help the yeast cell cycle clock tick: G2 cyclins transcriptionally activate G2 cyclins and repress G1 cyclins. Cell 74: 993–1007

Andrews BJ, Herskowitz I (1989a) The yeast SWI4 protein contains a motif present in developmental regulators and is part of a complex involved in cell-cycle-dependent transcription. Nature 342: 830–833

Andrews BJ, Herskowitz I (1989b) Identification of a DNA binding factor involved in cell-cycle control of the yeast *HO* gene. Cell 57: 21–29

Andrews BJ, Moore L (1992a) Mutational analysis of a DNA sequence involved in linking gene expression to the cell cycle. Biochem Cell Biol 70: 1073–1080

Andrews BJ, Moore LA (1992b) Interaction of the yeast Swi4 and Swi6 cell cycle regulatory proteins in vitro. Proc Natl Acad Sci USA 89: 11852–11856

Aves SJ, Durkacz BW, Carr A, Nurse P (1985) Cloning, sequencing and transcriptional control of the *Schizosaccharomyces pombe cdc10* start gene. EMBO J 4: 457–463

Ayte J, Leis JF, Herrera A, Tang E, Yang H, DeCaprio JA (1995) The *Schizosaccharomyces pombe* MBF complex requires heterodimerization for entry into S phase. Mol Cell Biol (in press)

Blackwell TK, Huang J, Ma A, Kretzner L, Alt FW, Eisenman RN, Weintraub H (1993) Binding of myc proteins to canonical and noncanonical DNA sequences. Mol Cell Biol 13: 5216–5224

Bork P (1993) Hundreds of ankyrin-like repeats in functionally diverse proteins: mobile modules that cross phyla horizontally. Prot Struct Funct Genet 17: 363–374

Breeden L, Mikesell G (1991) Cell cycle-specific expression of the SWI4 transcription factor is required for the cell cycle regulation of *HO* transcription. Genes Dev 5: 1183–1190

Breeden L, Mikesell G (1994) Three independent forms of regulation affect expression of *HO, CLN1* and *CLN2* during the cell cycle of *Saccharomyces cerevisiae*. Genetics 138: 1015–1024

Breeden L, Nasmyth K (1985) Regulation of the yeast *HO* gene. Cold Spring Harb Symp Quant Biol 50: 643–650

Breeden L, Nasmyth K (1987a) Cell cycle control of the yeast *HO* gene: *cis-* and*trans*-acting regulators. Cell 48: 389–397

Breeden L, Nasmyth K (1987b) Similarity between cell-cycle genes of budding yeast and fission yeast and the *Notch* gene of *Drosophila*. Nature 329: 651–654

Bueno A, Russell P (1992) Dual functions of *CDC6*: a yeast protein required for DNA replication also inhibits nuclear division. EMBO J 11: 2167–2176

Caligiuri M, Beach D (1993) Sct1 functions in partnership with Cdc10 in a transcription complex that activates cell cycle START and inhibits differentiation. Cell 72: 607–619

Cross FR (1988) *DAF1*, a mutant gene affecting size control, pheromone arrest, and cell cycle kinetics of *Saccharomyces cerevisiae*. Mol Cell Biol 8: 4675–4684

Cross FR, Tinkelenberg AH (1991) A potential positive feedback loop controlling *CLN1* and *CLN2* gene expression at the start of the yeast cell cycle. Cell 65: 875–883

Cross FR, Hoek M, McKinney JD, Tinkelenberg AH (1994) Role of Swi4 in cell cycle regulation of *CLN2* expression. Mol Cell Biol 14: 4779–4787

Dahmus ME (1994) On the role of C-terminal domain RNA polymerase II in the transcription of pre-mRNA. In: Conaway RC, Conaway JW (eds) Transcription, mechanisms and regulation. Raven, New York, pp243–262

DeCaprio J, Furukawa Y, Ajchenbaum F, Griffin J, Livingston D (1992) The retinoblastoma-susceptibility gene product becomes phosphorylated in multiple stages during cell cycle entry and progression. Proc Natl Acad Sci USA 89: 1795–1798

Dirick L, Nasmyth K (1991) positive feedback in the activation of G1 cyclins in yeast. Nature 351: 754–757

Dirick L, Moll T, Auer H, Nasmyth K (1992) A central role for *SWI6* in modulating cell cycle Start-specific transcription in yeast. Nature 357: 508–513

Donovan JD, Toyn JH, Johnson AL, Johnson LH (1994) P40 [SDB25], a putative CDK inhibitor, has role in the M/G$_1$ transition in *Saccharomyces cerevisiae*. Genes Dev 8: 1640–1653

Endicott JA, Nurse P, Johnson LN (1994) Mutational analysis supports a structural model for the cell cycle protein kinase p34. Prot Eng 7: 243–253

Epstein CB, Cross FR (1992) CLB5: a novel B cyclin from budding yeast with a role in S phase. Genes Dev 6: 1695–1706

Espinoza FH, Ogas J, Herskowitz I, Morgan DO (1994) Cell cycle control by a complex of the cyclin HCS26 (PCL1) and the kinase PHO85. Science 266: 1388–1391

Feaver WJ, Svejstrup JQ, Henry NL, Kornberg RD (1994) Relationship of CDK-activating kinase and RNA polymerase II CTD kinase TFIIH/TFIIIK. Cell 79: 1103–1109

Fesquet D, Labbe J-C, Derancourt J, Capony J-P, Galas S, Girard F, Iorca T, Shuttleworth J, Doree M, Cavadore J-C (1993) The MO15 gene encodes the catalytic subunit of a protein kinase that activates cdc2 and other cyclin dependent kinases (CDKs) through phosphorylation of Thr161 and its homologues. EMBO J 12: 3111–3121

Fitch I, Dahmann C, Surana U, Amon A, Nasmyth K, Goetsch L, Byers B, Futcher B (1992) Characterization of four B-type cyclin genes of the budding yeast *Saccharomyces cerevisiae*. Mol Biol Cell 3: 805–818

Foster R, Mikesell GE, Breeden L (1993) Multiple Swi6-dependent *cis*-acting elements control *SWI4* transcription through the cell cycle. Mol Cell Biol 13: 3792–3801

Ghiara J, Richardson H, Sugimoto K, Henze M, Lew D, Wittenberg C, Reed S (1991) A cyclin B homolog in *S. cerevisiae*: chronic activation of the Cdc28 protein kinase by cyclin prevents exit from mitosis. Cell 65: 163–174

Gimeno CJ, Fink GR (1994) Induction of pseudohyphal growth by overexpression of *PHD1*, a *Saccharomyces cerevisiae* gene related to transcriptional regulators of fungal development. Mol Cell Biol 14: 2100–2112

Gordon CB, Campbell JL (1991) A cell cycle-responsive transcriptional control element and a negative control element in the gene encoding DNA polymerase a in *Saccharomyces cerevisiae*. Proc Natl Acad Sci USA 88: 6058–6062

Haber JE, Garvik B (1977) A new gene affecting the efficiency of mating type interconversions in homothallic strains of *S. cerevisiae*. Genetics 87: 33–50

Hartwell LH (1976) Sequential function of gene products relative to DNA synthesis in the yeast cell cycle. J Mol Biol 104: 803–817

Hiebert SW, Chellappan SP, Horowitz JM, Nevins JR (1992) The interaction of RB with E2F coincides with an inhibition of the transcriptional activity of E2F. Genes Dev 6: 177–185

Hofmann JFX, Beach D (1994) *cdt1* is an essential target of the Cdc10/Sct1 transcription factor: requirement for DNA replication and inhibition of mitosis. EMBO J 13: 425–434

Hunter T, Karin M (1992) The regulation of transcription by phosphorylation. Cell 70: 375–387

Inoue, J, Kerr LD, Rashid D, Davis N, Bose HR Jr, Verma IM (1992) Direct association of pp40/kBb with Rel/NF-kB proteins: role of ankyrin repeats in the inhibition of DNA binding activity. Proc Natl Acad Sci USA 89: 4333–4337

Ivey-Hoyle H, Conroy R, Huber H, Goodhart P, Oliff A, Heimbrook D (1993) Cloning and characterization of E2F, a novel protein with the biochemical properties of transcription factor E2F. Mol Cell Biol 13: 7802–7812

Johnston LH, Lowndes NF (1992) Cell cycle control of DNA synthesis in budding yeast. Nucleic Acids Res 20: 2403–2410

Johnston LH, Thomas AP (1982) The isolation of new DNA synthesis mutants in the yeast *Saccharomyces cerevisiae*. Mol Gen Genet 186: 439–444

Kaffman A, Herskowitz I, Tijan R, O'Shea EK (1994) Phosphorylation of the transcription factor PHO4 by a cyclin-CDK complex, PHO80-PHO85. Science 263: 1153–1156

Kelly TJ, Martin GS, Forsburg SL, Stephen RJ, Russo A, Nurse P (1993) The fission yeast *cdc18*[+] gene product couples S phase to START and mitosis. Cell 74: 371–382

Koch C, Moll T, Neuberg M, Ahorn H, Nasmyth K (1993) A role for the transcription factors Mbp1 and Swi4 in progression from G1 to S phase. Science 261: 1551–1557

Kostriken R, Strathern JN, Klar AJS, Hicks JB, Heffron F (1983) A site specific nuclease essential for mating type switching in *S. cerevisiae*. Cell 35: 167–174

La Thangue NB (1994) DRTF1/E2F: an expanding family of heterodimeric transcription factors implicated in cell-cycle control. TIBS 19: 108–114

La Thangue NB, Taylor WR (1993) A structural similarity between mammalian and yeast transcription factors for cell-cycle-regulated genes. Trends Cell Biol 3: 75–76

LaMarco K, Thompson CC, Byers BP, Walton EM, McKnight SL (1991) Identification of Ets- and notch-related subunits in GA binding protein. Science 253: 789–792

Landschulz WH, Johnson PF, McKnight SL (1988) The leucine zipper: a hypothetical structure common to a new class of DNA binding proteins. Science 240: 1759–1764

Lees J, Saito M, Vidal M, Valentine M, Look T, Harlow E, Dyson N, Helin K (1993) The retinoblastoma protein binds to a family of E2F transcription factors. Mol Cell Biol 13: 7813–7825

Lowndes NT, Johnson AL, Johnston LH (1991) Coordination of expression of DNA synthesis genes in budding yeast by a cell-cycle regulated *trans* factor. Nature 350: 247–248

Lowndes NF, McInerny CJ, Johnson AL, Fantes PA, Johnston LH (1992a) Control of DNA synthesis genes in fission yeast by the cell-cycle gene *cdc10+*. Nature 355: 449–453

Lowndes NF, Johnson AL, Breeden L, Johnston LH (1992b) SWI6 protein is required for transcription of the periodically expressed DNA synthesis genes in budding yeast. Nature 357: 505–508

Marini NJ, Reed SI (1992) Direct induction of G_1-specific transcripts following reactivation of the Cdc28 kinase in the absence of de novo protein synthesis. Genes Dev 6: 557–567

Marks J, Fankhauser C, Reymond A, Simanis V (1993) Cytoskeletal and DNA structure abnormalities result from bypass of requirement for the *cdc10* start gene in fission yeast *Schizosaccharomyces pombe*. J Cell Sci 101: 517–528

McIntosh EM, Atkinson T, Storms RK, Smith M (1991) Characterization of a short, *cis*-acting DNA sequence which conveys cell cycle state-dependent transcription in *Saccharomyces cerevisiae*. Mol Cell Biol 11: 329–337

Mean AL, Slanksy JE, McMahon SL, Knuth MW, Farnham PJ (1992) The HIP binding site is required for growth regulation of the dihydrofolate reductase promoter. Mol Cell Biol 12: 1054–1063

Measday V, Moore L, Ogas J, Tyers M, Andrews B (1994) The PCL2 (ORFD)-PH085 cyclin-dependent kinase complex: A cell cycle regulator in yeast. Science 266: 1391–1395

Mendenhall MD (1993) An inhibitor of p34[CDC28] protein kinase activity from *Saccharomyces cerevisiae*. Science 259: 216–219

Mendenhall MD, Jones CA, Reed SI (1987) Dual regulation of the yeast *CDC28*-p40 protein kinase complex: Cell cycle, pheromone, and nutrient limitation effects. Cell 50: 927–935

Merrill GF, Morgan BA, Lowndes NF, Johnston LH (1992) DNA synthesis control of yeast: an evolutionarily conserved mechasnism for regulating DNA synthesis genes? Bioessays 14: 823–830

Miller KY, Toennis TM, Adams TH, Miller BL (1991) Isolation and transcriptional characterization of a morphological modifier: the *Aspergillus nidulans* stunted (*stuA)* gene. Mol Gen Genet 277: 285–292

Miller KY, Wu J, Miller BL (1992) StuA is required for cell pattern formation in *Aspergillus*. Genes Dev 6: 1770–1782

Miyamoto M, Tanaka K, Okayama H (1994) res2+, a new member of the *cdc10+/SWI4* family, controls the start of mitotic and meiotic cycle in fission yeast. EMBO J 13: 1873–1880

Moll T, Dirick L, Auer H, Bonkovsky J, Nasmyth K (1992) SWI6 is a regulatory subunit of two different cell cycle START-dependent transcription factors in *Saccharomyces cerevisiae*: J Cell Sci 16: 87–96

Moreno S, Nurse P (1990) Substrates for p34[cdc2]: In vivo veratis? Cell 61: 549–551

Nash R, Tokiwa G, Anand S, Erickson K, Futcher AB (1988) The *WHI1* gene of *Saccharomyces cerevisiae* tethers cell division to cell size and is a cyclin homolog. EMBO J 7: 4335–4346

Nasmyth K (1983) Molecular analysis of cell lineage. Nature 302: 670–676

Nasmyth K (1985a) At least 1400 base pairs of 5' flanking DNA is required for the correct expression of the *HO* gene in yeast. Cell 42: 213–223

Nasmyth K (1985b) A repetitive DNA sequence that confers cell cycle START (CDC28)-dependent transcription of the *HO* gene in yeast. Cell 42: 225–235

Nasmyth K, Dirick L (1991) The role of SWI4 and SWI6 in the activity of G_1 cyclins in yeast. Cell 66: 995–1013

Nasmyth K, Adolf G, Lydall D, Seddon A (1990) The identification of a second cell cycle control on the *HO* promoter in yeast: cell cycle regulation of SWI5 nuclear entry. Cell 62: 631–647

Nurse P, Bissett Y (1981) Gene required in G_1 for commitment to the cell cycle and in G_2 for control of mitosis in fission yeast. Nature 292: 558–560

Nurse P, Thuriaux P, Nasmyth K (1976) Genetic control of the cell division cycle in the fission yeast *S. pombe*. Mol Gen Genet 146: 167–178

Obara-Ishihara T, Okayama H (1994) A B-type cyclin negatively regulates conjugation via interacting with cell cycle 'start' genes in fission yeast. EMBO J 13: 1863–1872

Ogas J, Andrews BJ, Herskowitz I (1991) Transcriptional activation of *CLN1, CLN2*, and a putative new G_1 cyclin (*HCS26*) by SWI4, a positive regulator or G_1-specific transcription. Cell 66: 1015–1026

Ord RW, McIntosh EM, Lee L, Poon PP, Storms RK (1988) Multiple elements regulate expression of the cell cycle-regulated thymidylate synthase gene of *Saccharomyces cerevisiae*. Curr Genet 14: 363–373

Partridge J, Breeden L (to be published) Multiple regulatory elements control CLN1 transcription during the cell cycle. EMBO J

Peter M, Herskowitz I (1994) Direct inhibition of the yeast cyclin-dependent kinase Cdc28-Cln by Far1.Science 265: 1228–1231

Poon RYC, Yamashita K, Adamczewski JP, Hunt T, Shuttleworth J (1993) The cdc2-related protein $p40^{MO15}$ is the catalytic subunit of a protein kinase that can activate $p33^{cdk2}$ and $p34^{cdc2}$. EMBO J 12: 3123–3132

Primig M, Sockanathan S, Auer H, Nasmyth K (1992) Anatomy of a transcription factor important for the Start of the cell cycle in *Saccharomyces cerevisiae*. Nature 358: 593–597

Reid B, Hartwell LH (1977) Regulation of mating in the cell cycle of *Saccharomyces cerevisiae*. J Cell Biol 75: 355–365

Reymond A, Schmidt S, Simanis V (1992) Mutations in the *cdc10* start gene of *Schizosaccharomyces pombe* implicate the region of homology between *cdc10* and *SWI6* as important for $p85^{cdc10}$ function. Mol Gen Genet 234: 449–456

Reymond A, Marks J, Simanis V (1993) The activity of *S. pombe* DSC-1-like factor is cell cycle regulated and dependent on the activity of $p34^{cdc2}$. EMBO J 12: 4325–4334

Richardson H, Lew DJ, Henze M, Sugimoto K, Reed SI (1992) Cyclin-B homologs in *S. cerevisiae* function in S phase and in G_2. Genes 6: 2021–2034

Richardson HE, Wittenberg C, Cross F, Reed SI (1989) An essential G_1 function for cyclin-like proteins in yeast. Cell 59: 1127–1133

Roy R, Adamczewski JP, Seroz T, Vermeulen W, Tassan J-P, Schaeffer L, Nigg EA, Hoeijmakers JHJ, Egly J-M (1994) The MO15 cell cycle kinase is associated with the TFIIH transcription-DNA repair factor. Cell 79: 1093–1101

Schwob E, Nasmyth K (1993) CLB5 and CLB6, a new pair of B cyclins involved in DNA replication in *Saccharomyces cerevisiae*. Genes Dev 7: 1160–1175

Schwob E, Bohm T, Mendenhall MD, Nasmyth K (1994) The B-type cyclin kinase inhibitor $p40^{SIC1}$ controls the G_1 to S transition in *S. cerevisiae*. Cell 79: 233–244

Shan B, Zhu, X, Chen P, Durfee T, Yang Y, Sharp D, Lee W (1992) Molecular cloning of cellular genes encoding retinoblastoma-associated proteins: identification of a gene with properties of the transcription factor E2F. Mol Cell Biol 12: 5620–5631

Sidorova J, Breeden L (1993) Analysis of the SWI4/SWI6 protein complex, which directs G_1/S-specific transcription in *Saccharomyces cerevisiae*: Mol Cell Biol 13: 1069–1077

Sidorova J, Breeden L (to be published) Phosphorylation of Swi6 controls its nuclear localization but doesn't affect cell cycle regulated transcription. Mol Biol Cell

Simanis V, Nurse P (1989) Characterization of the fission yest *cdc10*⁺ protein that is required for commitment to the cell cycle. J Cell Sci 92: 51–56

Simon M, Seraphin B, Gerard F (1986)*KIN28*, a yeast gene coding for a putative protein kinase homologous to *CDC28*. EMBO J 5: 2697–2701

Sipiczki M (1989) Taxonomy and phylogenesis. In: Nasim A, Johnson, BF, Young P (eds) Molecular biology of the fission yeast. Academic, New York, pp 431–452

Stern M, Jenson R, Herskowitz I (1984) Five *SWI* genes are required for expression of the *HO* gene in yeast. J Mol Biol 178: 853–868

Strathern JN, Herskowitz I (1979) Assymetry and directionality in production of new cell types during clonal growth: the switching pattern of homothallic yeast. Cell 17: 371–381

Stuart D, Wittenberg C (1994) Cell cycle-dependent transcription of *CLN2* is conferred by multiple distinct *cis*-acting regulatory elements. Mol Cell Biol 14: 4788–4801

Taba MRM, Muroff I, Lydall D, Tebb G, Nasmyth K (1991) Changes in a SWI4, 6-DNA-binding complex occur at the time of *HO* gene activation in yeast. Genes Dev 5: 2000–2013

Tanaka K, Okazaki K, Okazaki N, Ueda T, Sugiyama A, Nojima H, Okayama H (1992) A new *cdc* gene required for S phase entry of *Schizosaccharomyces pombe* encodes a protein similar to the *cdc10*⁺ and *SWI4* gene products. EMBO J 11: 4923–4932

Tyers M, Tokiwa G, Futcher B (1993) Comparison of the *Saccharomyces cerevisiae* G$_1$ cyclins: Cln3 may be an upstream activator of Cln1, Cln2 and other cyclins. EMBO J 12: 1955–1968

Valay JG, Simon M, Faye G (1993) The Kin28 protein kinase is associated with a cyclin in *Saccharomyces cerevisiae*. J Mol Biol 234: 307–310

Ward MP, Garrett S (1994) Suppression of a yeast cyclic AMP-dependent protein kinase defect by overexpression of *SOK1*, a yeast gene exhibiting sequence similarity to a developmentally regulated mouse gene. Mol Cell Biol 14: 5619–5627

Weinert TA, Kiser GL, Hartwell LH (1994) Mitotic checkpoint genes in budding yeast and the dependence of mitosis on DNA replication and repair. Genes Dev 8: 652–665

White JHM, Green SR, Barker DG, Dumas LB, Johnston LH (1987) The *CDC8* transcript is cell cycle regulated in yeast and is expressed with *CDC9* and *CDC21* at a point preceding histone transcription. Exp Cell Res 171: 223–231

Wittenberg C, Sugimoto K, Reed SI (1990) G$_1$-specific cyclins of *S. cerevisiae*: cell cycle periodicity, regulation by mating pheromone, and association with the p34^{CDC28} protein kinase. Cell 62: 225–237

Zheng P, Fay DS, Burton J, Xiao H, pinkham JL, Stern DF (1993) *SPK1* is an essential S-phase-specific gene of *Saccharomyces cerevisiae* that encodes a nuclear serine/threonine/tyrosione kinase. Mol Cell Biol 13: 5829–5842

Zhou C, Jong AY (1993) Mutation analysis of *Saccharomyces cerevisiae CDC6* promoter: defining its UAS domain and cell cycle regulating element. DNA Cell Biol 4: 363–370

Zhu Y, Takeda T, Nasmyth K, Jones N (1994) *pct1*$^+$, which encodes a new DNA-binding partner of p85P^{cdc10}, is required for meiosis in the fission yeast *Schizosaccharomyces pombe*. Genes Dev 8: 885–898

Conclusions and Future Directions

P.J. FARNHAM

1 Unanswered Questions

The decision to proliferate or to enter a non replicating state is one that a cell must continually reassess. This decision is based in part upon the environmental cues encountered, such as the levels of growth factors, the presence of chemotherapeutic drugs, and infection by viruses. Investigations into the molecular mechanisms by which cells control their proliferative response have focused on positive effectors such as cyclins and viral oncoproteins, negative effectors such as tumor suppressor proteins, and signal transduction pathways leading to the transcriptional activation of various genes in particular stages of the cell cycle. Each of these approaches has implicated E2F in the control of cell growth. Although progress towards understanding the E2F gene family has been rapid, and great advances have been made in the last several years, questions concerning the different family members and their role in cell growth control still remain. Several specific questions that remain to be addressed are discussed below.

1.1 Do Individual E2F Family Members Have Distinct Functions?

As described by SLANSKY and FARNHAM (this volume), it has not been satisfactorily determined which E2F family member activates which specific cellular promoter. The differences in expression profiles of the various E2F family members suggest that they regulate distinct genes. To date no individual preferences for recognition sequences have been observed; all E2Fs can bind to the consensus site. A

McArdle Laboratory for Cancer Research, 1400 University Avenue, University of Wisconsin Medical School, Madison, WI 53706, USA consideration of the regulation of G_1/S phase transcription in yeast highlights the fact that simple sequence inspection cannot substitute for direct experimentation in the analysis of transcriptional regulation. For example, the CLN1 promoter in *Saccharomyces cerevisiae* is regulated by an MCB element (see BREEDEN, this volume). Although other MCB elements are bound by MBP1/Swi6-protein complexes, the MCB element in the CLN1 promoter is bound by Swi4/Swi6-protein complexes (which normally bind to SCB elements). This suggests that there are determinants other than the core sequence elements that control which G_1/S phase regulator binds to and activates a particular promoter in yeast. Although most E2F sites show increased binding activity in S phase extracts, the site from the mammalian insulin-like growth factor type 1 (IGF-1) promoter does not (see SLANSKY and FARNHAM, this volume). Perhaps the inability to detect binding to the E2F site from the IGF-1 promoter in S phase extracts is an example of how flanking sequences can determine protein binding specificity (PORCU et al. 1994).

All approaches utilized to date to determine which E2F regulates a particular cellular gene have distinct drawbacks that weaken interpretation of the results. For example, the use of in vitro systems or reporter-gene assays does not allow the contribution of chromosomal location and long-range structure of a promoter to be considered. The importance of chromosomal location is suggested by studies of the G_1/S phase regulation of the *HO* promoter in yeast. Overexpression of Swi4, the transcription factor thought to regulate the expression of the *HO* gene, does not abolish the periodicity of the transcriptional activity from the *HO* promoter. This suggests that another factor may also function to prevent *HO* promoter activity in G_1 cells. If multiple forms of repression exist to keep proliferation-related promoters off in quiescent mammalian cells, results using small promoter fragments in reporter assays may not reproduce the response of the endogenous promoter. For example, as noted by SLANSKY and FARNHAM (this volume), activity from a *dhfr* promoter fragment used in reporter assays is increased upon expression of E2F1. However, cells stably expressing E2F1 do not have elevated levels of endogenous *dhfr* mRNA (J.R. NEVINS, personal communication; SINGH et al. 1994). Perhaps the endogenous *dhfr* promoter is also regulated by quiescent-specific transcriptional repressors whose binding sites lie outside of the promoter region used in reporter assays. Studies of the mouse thymidine kinase promoter have shown that although an E2F complex does bind to the promoter, a nearby element that has a different sequence also displays serum-regulated changes in binding activity. This element, termed Yi, contains cyclin D1/cdk2 and a 110-kDa DNA-binding subunit. A cDNA has been cloned that encodes a protein that can bind to the Yi element. Although this protein has a different amino acid sequence than E2F, it can bind to bacterially expressed retinoblastoma (Rb) protein, suggesting that binding of a complex of the 110-kDa protein and Rb might assist E2F in repressing the thymidine kinase promoter in G_0 cells (DOU et al. 1994).

Overexpression of different E2F family members in a cell may give information about what family member *could* regulate a specific target gene, but does not address the question as to what family member *does* activate that gene under normal growth conditions. Two alternative approaches towards determining which E2F family member activates a particular cellular promoter are described below.

The first approach would be individually to remove or reduce the levels of an E2F family member in a cell using methods such as antibodies, ribozymes, or antisense mRNAs. Unchanged levels of a specific cellular mRNA after removal of an E2F family member would indicate that the particular E2F is not essential for activation of that promoter. Since the E2F family is thought to be critical in growth control, an inducible or transient system is required so that removal of a family member does not impose selective growth conditions on the cells. To date no studies have injected or transfected family member specific antibodies into cells. Initial attempts to target cellular *E2F1* mRNA with ribozymes have not been successful (J.E. Slansky and P.J. Farnham, unpublished data); perhaps an alternative approach using selection to find the most active anti-E2F1 ribozyme would be more effective (LIEBER and STRAUSS 1995). Expression of a dexamethazone-inducible antisense *E2F1* has been shown to greatly reduce the amount of *E2F1* protein in T98G glioblastoma cells (SALA et al. 1994). Although the levels of cellular mRNAs thought to be regulated by the E2F family were not examined in this study, the successful reduction of E2F1 protein suggests that this approach may be the most fruitful. For each of the described methods, it is important to also investigate the effects of removal of an E2F on the levels of the other family members. If removal of a single E2F alters the amounts of the other E2Fs, these experiments may not be instructive.

Another approach towards assigning specific E2F family members to specific promoters would be based on immunoprecipitation assays. Studies of heat shock promoters have employed a method in which cells or nuclei are UV-irradiated to cross-link proteins covalently to the DNA (GILMOUR and LIS 1985, 1986; WALTER et al. 1994). The DNA is then isolated and digested with restriction enzymes that cut in known regions of the gene(s) of interest. After immunoprecipitation with antibodies against the transcription factor, the DNA is cleared of protein and used in a Southern analysis with probes for specific promoters. Family member specificity could be ascertained if the DNA from some but not all E2F site containing promoters were immunopreciptated with antibodies raised against individual family members.

In summary, no investigations performed to date can clearly demonstrate a unique role of an individual E2F family member in the activation of a particular cellular promoter or in the initiation of a cellular response.

1.2 Is E2F Activity Critical for Cell Proliferation?

As described by ADAMS and KAELIN (this volume) overexpression of E2F1 or E2F4 can drive certain quiescent cell types into S phase. These experiments suggest that the expression of E2F-activated genes can substitute for growth factor stimulation. Although the products of E2F target genes such as DHFR and DNA polymerase-α are obviously required for cell proliferation, it is unlikely that increased levels of these proteins provide a complete mitogenic signal. However, other E2F target genes such as b-*myb* and c-*myc* are transcription factors; perhaps activation of these genes provides the necessary branching of the E2F-initiated signal that is required to activate all genes needed for DNA synthesis to occur.

Increased E2F activity is not always sufficient to overcome a cell cycle arrest. Both Rb and p107 can repress E2F-mediated transcription and they can both cause SAOS-2 cells to arrest in the G_1 phase of the cell cycle. However, overexpression of E2F1 can reverse the Rb-induced cell cycle arrest, but not the p107-induced arrest (Zhu et al. 1993). This difference could be attributed to the fact that E2F1 preferentially binds to Rb, not p107. Perhaps overexpression of E2F4, a p107-specific E2F, would relieve the p107 but not the Rb block. Although this experiment has not yet been performed, a cell cycle arrest induced by a similar protein, p130, can be relieved by expression of E2F4 (Vairo et al. 1995). In fitting with the preference of E2F4 to bind to p107 or p130 but not Rb, E2F4 cannot effectively relieve an Rb-induced cell cycle arrest (Vairo et al. 1995). It is also possible that p107 inhibits more than just E2F-mediated transcription so that overexpression of an E2F would not completely bypass a p107 block. In support of this hypothesis, a mutant p107 protein that cannot interact with E2F can still repress growth (Smith and Nevins 1995). It has recently been demonstrated that p107 can bind to the transactivation domain of the c-Myc protein in vivo (Beijersbergen et al. 1994), and that the binding of p107 to c-Myc causes inhibition of Myc-mediated transactivation. Furthermore, expression of c-Myc can partially relieve a p107-, but not an Rb-induced growth arrest. Based on the observations that E2F and c-Myc may activate different G_1/S phase specific promoters (Miltenberger et al. 1995), it would be of interest to determine whether coexpression of c-Myc plus E2F1 elicits a greater proliferative response.

Although the link between the E2F family and transcriptional regulation of genes required for DNA replication is strong (see Slansky and Farnham, this volume), the *necessity* for E2F activity in cell proliferation has not been well studied. To determine the role of E2F in these processes, it is necessary to remove the activity of the E2F family members (individually or collectively) from a cell. Although several groups have begun to use knock-out strategies to make mice lacking different E2F family members, no results are yet available. Knock-out strategies, of course, have several drawbacks. If a family member is essential, removal may be lethal to the developing organism. Alternatively, if family members are redundant, the effects of individually removing a particular family member may be slight. For example, removal of MBP1 or Swi4 activity from a yeast cell does not produce a lethal phenotype; however, the double mutation is lethal. It is possible that there is functional redundancy of these two DNA-binding proteins in cells such that loss of one allows the other to compensate by activating all of the G_1/S phase specific promoters. Perhaps investigators of the E2F gene family may find that several (or all?) of the E2F genes need to be mutated before a phenotype can be observed. As described above, attempts to remove E2F family members individually using ribozymes have not been successful, and only one study has been published using antisense E2F mRNA. Expression of a dexamethazone-inducible antisense E2F1 in T98G glioblastoma cells delays the onset of the first G_2 phase after quiescent cells are stimulated to begin proliferation (SALA et al, 1994). These results suggest that E2F1-mediated transcription is required to rapidly pass through S phase and are consistent with the G_1/S phase expression pattern of E2F1 and with the idea that genes required for DNA synthesis, such as *dhfr*, DNA polymerase-α, and thymidine kinase, are regulated by E2F1. However, the time points of the experiment were not extended long enough to determine whether the cells eventually enter G_2 and M phases, nor were the effects of the antisense E2F1 examined in growing cells.

Several studies have examined the effects of a general inhibition of E2F activity on cell proliferation. One set of studies achieved inhibition of E2F activity by overexpression of Rb or p107. In certain cell types (such as T98G glioblastoma cells), overexpression of either Rb or p107 results in reduced E2F-mediated transcription and a higher percentage of G_1 cells (ZHU et al. 1993). In contrast, repression of E2F activity in C33A cells by overexpression of Rb does not lead to an arrest of cell proliferation (ZHU et al. 1993), suggesting that in some tumor cells E2F-mediated transcription is not required for progression through the cell cycle. E2F activity has also been inhibited using dominant negative proteins such as a mutant E2F1 which lacks a transactivation domain. This protein should dimerize with the normal DP partner, bind to cellular promoters, and block wild-type E2F from binding to the DNA. This dominant negative E2F1 was shown to inhibit the ability of E1A to induce DNA synthesis of quiescent BALB/c 3T3 cells (DOBROWOLSKI et al. 1994). Although these results suggest that E2F activity is required for virally mediated oncogenesis, this hypothesis has not yet been examined directly.

In summary, investigations into the necessity and sufficiency of E2F activity in the control of normal and neoplastic growth have just begun: all of these studies have focused on the role of E2F in cell culture systems. Current investigations suggest that E2F does play a critical role in the G_1 to S phase transition. Further studies using transgenic animals and analysis of primary human tumors are required to determine whether E2F family members play a critical role in controlling the proliferative response in whole organisms.

1.3 Is E2F the Only Mediator of G_1/S Phase Transcription?

In some cell types, the consequence of overexpression of E2F1 is cell death, not cell proliferation, suggesting that E2F may not provide a complete mitogenic signal in these cases. It has been proposed that activation of some, but not all, of the necessary proliferation pathways leads to conflicting signals that trigger apoptosis. Are other transcriptional regulators needed in addition to the E2F gene family for a cell to successfully progress from G_1 into S phase?

As described by BREEDEN (this volume), the transcriptional regulation of genes required for the G_1 to S phase progression (i.e., passage through Start) in *S. cerevisiae* is controlled by transcription complexes that bind to two different elements. Start-specific transcription is regulated by two different DNA-binding proteins, Swi4 or MBP1, each of which forms a heterodimer with a common third protein, Swi6. The existence of two related, but different proteins that regulate G_1/S phase specific transcription in yeast raises the possibility that transcription factors related to but distinct from E2F may contribute to G_1/S phase transcription in mammalian cells. Studies of the human thymidine kinase and hamster *cad* promoters support this hypothesis.

Analyses of the human thymidine kinase promoter suggest that a protein that has a DNA-binding specificity similar to but distinct from E2F family members may bind to and regulate G_1/S phase specific transcription. Although mutation of a element (TCTCCCGC) that is a 7/8 match to a consensus E2F site reduced regulation, a consensus E2F site was not an efficient competitor for binding to the human thymidine kinase element (LI et al. 1993). The authors suggest that perhaps a protein related to E2F regulates human thymidine kinase. Alternatively, flanking sequences near the thymidine kinase E2F site could specify binding of an E2F family member that is different than the one that binds to the E2F site tested as a competitor. However, in support of the hypothesis that a related but distinct protein activates the thymidine kinase promoter, a new human gene has been cloned that is homologous to the E2F family of transcription factors (Y. VAISHNAV, personnal communication). The overall homology of this protein to E2F family members is quite low (35%); however, a higher homology can be found to the dimerization (65%) and Rb-binding domain (53%). It remains to be seen whether this new gene product displays E2F-like activities such as recognition of specific DNA sequences and transcriptional activation.

Other investigations have shown that sequence elements distinct from E2F sites can mediate G_1/S phase specific transcription. The *cad* gene encodes the first three steps in pyrimidine biosynthesis and is thus necessary for DNA replication. Transcription from the *cad* promoter is increased in late G_1, and the growth-responsive promoter element has been mapped to a consensus binding site for the c-Myc oncogene (MILTENBERGER et al. 1995). The c-Myc protein shares many similarities to the E2F gene family. It encodes a growth-regulated transcription factor that heterodimerizes with a constitutively present partner called Max (BLACKWOOD et al. 1992). As do the E2F family members, c-Myc also binds to DNA via a helix-loop-helix protein domain and interacts with pocket proteins such as p107. Although the Myc/Max heterodimer is a transcriptional activator, other Max-binding partners (such as Mad or MxiI) have been identified that create heterodimers that act to repress transcription (AYER and EISENMAN 1993; ZERVOS et al. 1993). Mad/Max heterodimers may serve to keep proliferation-specific promoters off in terminally differentiated cells. Although such partners for E2F1 or DP1 have not yet been demonstrated, future investigations may identify such E2F-related transcriptional repressors that function during differentiation.

In summary, it is unlikely that the E2F family is the sole regulator of the G_1 to S phase transition. The ability of E2F to drive cells into S phase may indicate that in certain cell culture systems, other signal transduction pathways may already be constitutively activated. Further analysis of the cooperation of E2F with other growth-related transcription factors is required before a thorough understanding of cell growth control is achieved.

2 Concluding Remarks

In less than 10 years we have progressed from a vague understanding that some cellular promoters are transcriptionally activated at the G_1/S phase boundary to the identification and characterization of seven members of a gene family that are responsible for at least part of this regulation. The identification of the E2F gene family has allowed a more complete understanding of how positive and negative regulators of cell growth function. For example, although it was clear that loss of (Rb) protein is associated with human neoplasias, it was not clear how this tumor suppressor protein inhibits cell growth. The discovery that Rb can repress the activity of E2F-mediated transcription and thus reduce the synthesis of proteins required for DNA replication provided insight into cell growth control. Although cyclins were known to be the growth-regulated component of multiple kinase complexes, it was not known what cellular proteins are targets of these kinases. The finding that both G_1 and S phase cyclin/kinase complexes regulate E2F activity has held to a model for cyclin function. The discovery that viral oncoproteins can also activate E2F-mediated transcription via release of E2F proteins from inhibitory complexes has also allowed a greater understanding of viral oncogenesis. Thus, studies of E2F have allowed an integrated understanding of many aspects of cell growth control.

If this gene family is of central importance in the control of the decision to enter S phase and begin DNA replication, it is likely that functional redundancy has been built into the system. Loss of the ability to make proteins required for DNA replication would be a lethal event. Perhaps the existence of such a large number of E2F family members allows a cell to suffer inactivating mutations in several genes without losing the ability to make proteins required for DNA replication. On the other hand, it is also essential to have precise regulation over the E2F transcription factors. Loss of the ability to keep these proteins inactive in quiescent or differentiated cells may lead to neoplasia. The best characterized E2F family member, E2F1, has been shown to regulated via protein-protein interactions (see COBRINIK, this volume), mRNA instability (R.J. Szakaly and P.J. Farnham, unpublished data), and transcriptional activation (HSIAO et al. 1994; JOHNSON et al. 1994; NEUMAN et al. 1994). Perhaps the existance of redundant control mechanisms allow a cell to suffer loss of one mechanism without loss of cell growth control. Based on studies to date, it is clear that the regulatory networks involving the E2F gene family are quite complicated. Future studies of how this gene family is regulated and how it regulates cell growth will provide further insight into the intricacies of cell proliferation.

Acknowledgments. We thank other contributors to this volume for their time and effort spent in providing an up-to-date analysis of the E2F gene family. We are also grateful to those investigators who allowed us to include their unpublished data throughout this volume; these open lines of communication foster an *esprit de corps* that allows rapid progress in understanding cell growth control. Finally, we thank Jackie Lees, Nick Dyson, Bob Weinberg, Joe Nevins, Paul Lambert, Ira Herskowitz, Jill Slansky, and Brenda Andrews who were willing to serve as readers of the various chapters; their efforts assured that we could provide a balanced and complete picture of the E2F gene family.

References

Ayer DE, Eisenman RN (1993) A switch from Myc: Max to Mad: Max heterocomplexes accompanies monocyte/macrophage differentiation. Genes Dev 7: 2110–2119

Beijersbergen RL, Hijmans Em, Zhu L, Bernards R (1994) Interaction of c-Myc with the pRb-related protein p107 results in inhibition of c-Myc-mediated transactivation. EMBO J 13: 4080–4086

Blackwood EM, Kretzner L, Eisenman RN (1992) Myc and Max function as a nucleoprotein complex. Curr Opin Genet Dev 2: 227–235

Dobrowolski SF, Stacey DW, Harter MW, Stine JT, Hiebert SW (1994) An E2F dominant negative mutant blocks E1A induced cell cycle progression. Oncogene 9: 2605–2612

Dou Q-P, Molnar G, Pardee AB (1994) Cyclin D1/CDK2 kinase is present in a G1 phase-specific protein complex Yl1 that binds to the mouse thymidine kinase gene promoter. Biochem Biophys Res Commun 205: 1859–1868

Gilmour DS, Lis JT (1985) In vivo interactions of RNA polymerase II with genes of Drosophila melanogaster. Mol Cell Biol 5: 2009–2018

Gilmour DS, Lis JT (1986) RNA polymerase II interacts with the promoter region of the noninduced hsp70 gene in Drosophila melanogaster cells. Mol Cell Biol 6: 3984–3989

Hsiao K-M, McMahon SL, Farnham PJ (1994) Multiple DNA elements are required for the growth regulation of the mouse E2F1 promoter. Genes Dev 8: 1526–1537

Johnson DG, Ohtani K, Nevins JR (1994) Autoregulatory control of E2F1 expression in response to positive and negative regulators of cell cycle progression. Genes Dev 8: 1514–1525

Li L, Naeve GS, Lee AS (1993) Temporal regulation of cyclin A-p107 and p33cdk2 complexes binding to human thymidine kinase promoter element important for G_1-S Phase transcriptional regulation. Proc Natl Acad Sci USA 90: 3554–3558

Lieber A, Strauss M (1995) Selection of efficient cleavage sites in target RNAs by using a ribozyme expression library. Mol Cell Biol 15: 540–551

Miltenberger RJ, Sukow K, Farnham PJ (1995) An E box-mediated increase in cad transcription at the G_1/S-phase boundary is suppressed by inhibitory c-Myc mutants. Mol Cell Biol 15: 2527–2535

Neuman E, Flemington EK, Sellers WR, Kaelin WG Jr (1994) Transcription of the E2F-1 gene is rendered cell cycle dependent by E2F DNA-binding sites within its promoter. Mol Cell Biol 14: 6607–6615

Porcu P, Grana X, Li S, Swantek J, De Luca A, Giordano A, Baserga R (1994) An E2F binding sequence negatively regulates the response of the insulin-like growth factor 1 (IGF-I) promoter to simian virus 40 T antigen and to serum. Oncogene 9: 2125–2134

Sala A, Nicolaides Nc, Engelhard A, Bellon T, Lawe DC, Arnold A, Grana X, Giordano A, Calabretta B (1994) Correlation between E2F-1 requirement in S phase and E2F-1 transactivation of cell cycle related genes in human cells. Cancer Res 54: 1402–1406

Singh P, Wong SH, Hong W (1994) Overexpression of E2F-1 in rat embryo fibroblasts leads to neoplastic transformation. EMBO J 13: 3329–3338

Smith EJ, Nevins JR (1995) The Rb-related p107 protein can suppress E2F function independently of binding to cyclin A/cdk2. Mol Cell Biol 15: 338–344

Vairo G, Livingston DM, Ginsberg D (1995) Functional interaction between E2F4 and p130: evidence for distinct mechanisms underlying growth suppression by different Rb family members. Genes Dev 9: 869–881

Walter J, Dever CA, Biggin MD (1994) Two homeo domain proteins bind with similar specificity to a wide range of DNA sites in Drosophila embryos. Genes Dev 8: 1678–4692

Zervos AS, Gyuris J, Brent R (1993) Mxi1, a protein that specifically interacts with Max to bind Myc-Max recognition sites. Cell 72: 223–232

Zhu L, van den Heuvel S, Helin K, Fattaey A, Ewen M, Livingston D, Dyson N, Harlow E (1993) Inhibition of cell proliferation by p107, a relative of the retinoblastoma protein. Genes Dev 7: 1111–1125

Subject Index

Current Topics in Microbiology and Immunology

Volumes published since 1989 (and still available)

Vol. 186: **zur Hausen, Harald (Ed.):** Human Pathogenic Papillomaviruses. 1994. 37 figs. XIII, 274 pp. ISBN 3-540-57193-0

Vol. 187: **Rupprecht, Charles E.; Dietzschold, Bernhard; Koprowski, Hilary (Eds.):** Lyssaviruses. 1994. 50 figs. IX, 352 pp. ISBN 3-540-57194-9

Vol. 188: **Letvin, Norman L.; Desrosiers, Ronald C. (Eds.):** Simian Immunodeficiency Virus. 1994. 37 figs. X, 240 pp. ISBN 3-540-57274-0

Vol. 189: **Oldstone, Michael B. A. (Ed.):** Cytotoxic T-Lymphocytes in Human Viral and Malaria Infections. 1994. 37 figs. IX, 210 pp. ISBN 3-540-57259-7

Vol. 190: **Koprowski, Hilary; Lipkin, W. Ian (Eds.):** Borna Disease. 1995. 33 figs. IX, 134 pp. ISBN 3-540-57388-7

Vol. 191: **ter Meulen, Volker; Billeter, Martin A. (Eds.):** Measles Virus. 1995. 23 figs. IX, 196 pp. ISBN 3-540-57389-5

Vol. 192: **Dangl, Jeffrey L. (Ed.):** Bacterial Pathogenesis of Plants and Animals. 1994. 41 figs. IX, 343 pp. ISBN 3-540-57391-7

Vol. 193: **Chen, Irvin S. Y.; Koprowski, Hilary; Srinivasan, Alagarsamy; Vogt, Peter K. (Eds.):** Transacting Functions of Human Retroviruses. 1995. 49 figs. IX, 240 pp. ISBN 3-540-57901-X

Vol. 194: **Potter, Michael; Melchers, Fritz (Eds.):** Mechanisms in B-cell Neoplasia. 1995. 152 figs. XXV, 458 pp. ISBN 3-540-58447-1

Vol. 195: **Montecucco, Cesare (Ed.):** Clostridial Neurotoxins. 1995. 28 figs. XI., 278 pp. ISBN 3-540-58452-8

Vol. 196: **Koprowski, Hilary; Maeda, Hiroshi (Eds.):** The Role of Nitric Oxide in Physiology and Pathophysiology. 1995. 21 figs. IX, 90 pp. ISBN 3-540-58214-2

Vol. 197: **Meyer, Peter (Ed.):** Gene Silencing in Higher Plants and Related Phenomena in Other Eukaryotes. 1995. 17 figs. IX, 232 pp. ISBN 3-540-58236-3

Vol. 198: **Griffiths, Gillian M.; Tschopp, Jürg (Eds.):** Pathways for Cytolysis. 1995. 45 figs. IX, 224 pp. ISBN 3-540-58725-X

Vol. 199/I: **Doerfler, Walter; Böhm, Petra (Eds.):** The Molecular Repertoire of Adenoviruses I. 1995. 51 figs. XIII, 280 pp. ISBN 3-540-58828-0

Vol. 199/II: **Doerfler, Walter; Böhm, Petra (Eds.):** The Molecular Repertoire of Adenoviruses II. 1995. 36 figs. XIII, 278 pp. ISBN 3-540-58829-9

Vol. 199/III: **Doerfler, Walter; Böhm, Petra (Eds.):** The Molecular Repertoire of Adenoviruses III. 1995. 51 figs. XIII, 310 pp. ISBN 3-540-58987-2

Vol. 200: **Kroemer, Guido; Martinez-A., Carlos (Eds.):** Apoptosis in Immunology. 1995. 14 figs. XI, 242 pp. ISBN 3-540-58756-X

Vol. 201: **Kosco-Vilbois, Marie H. (Ed.):** An Antigen Depository of the Immune System: Follicular Dendritic Cells. 1995. 39 figs. IX, 209 pp. ISBN 3-540-59013-7

Vol. 202: **Oldstone, Michael B. A.; Vitković, Ljubiša (Eds.):** HIV and Dementia. 1995. 40 figs. XIII, 279 pp. ISBN 3-540-59117-6

Vol. 203: **Sarnow, Peter (Ed.):** Cap-Independent Translation. 1995. 31 figs. XI, 183 pp. ISBN 3-540-59121-4

Vol. 204: **Saedler, Heinz; Gierl, Alfons (Eds.):** Transposable Elements. 1995. 42 figs. IX, 234 pp. ISBN 3-540-59342-X

Vol. 205: **Littman, Dan. R. (Ed.):** The CD4 Molecule. 1995. 29 figs. XIII, 182 pp. ISBN 3-540-59344-6

Vol. 206: **Chisari, Francis V.; Oldstone, Michael B. A. (Eds.):** Transgenic Models of Human Viral and Immunological Disease. 1995. 53 figs. XII, 350 pp. ISBN 3-540-59341-1

Vol. 207: **Prusiner, Stanley B. (Ed.):** Prions Prions Prions. 1995. 42 figs. VI, 180 pp. ISBN 3-540-59343-8

Springer-Verlag
and the Environment

We at Springer-Verlag firmly believe that an international science publisher has a special obligation to the environment, and our corporate policies consistently reflect this conviction.

We also expect our business partners – paper mills, printers, packaging manufacturers, etc. – to commit themselves to using environmentally friendly materials and production processes.

The paper in this book is made from low- or no-chlorine pulp and is acid free, in conformance with international standards for paper permanency.